Home Electrics

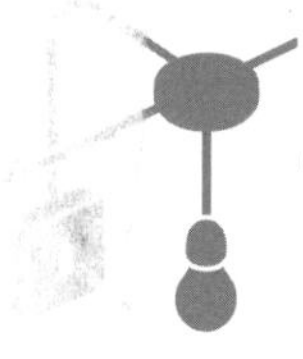

THE COMPLETE HANDBOOK

JULIAN BRIDGEWATER

In 2008, the 17th edition (BS 7671.2008) of wiring regulations was published by the IEE. These regulations should be consulted in conjunction with this book.

DEDICATION

To my wonderful wife, who has graciously made me many cups of tea while I pull the house to bits and 'DIY it' back together... eventually!

This edition published in 2009 by
New Holland Publishers (UK) Ltd
Garfield House, 86–88 Edgware Road
London W2 2EA
United Kingdom
London • Cape Town • Sydney • Auckland
www.newhollandpublishers.com

6 8 10 9 7

ISBN 978 1 84773 436 5

Editorial Direction: Rosemary Wilkinson
Editors: Gareth Jones, Karen Hemingway and Fiona Biggs
Proof-reader: Jacky Jackson
Design: AG&G Books
Illustrations: Peters & Zabransky (UK) Ltd
Cover photograph: Ian Parsons

Printed and bound by Kyodo Nation Printing Services, Thailand

NOTE

The author and publishers have made every effort to ensure that all instructions given in this book are safe and accurate, but they cannot accept liability for any resulting injuries or loss or damage to either property or person, whether direct or consequential and howsoever arising.

Contents

Introduction

Of all the DIY activities in the home, electrical work - mending and replacing electrical systems round the home - is the activity that is considered the most dangerous. If you were to walk up your local high street and do a mini survey, you would find that, while most people wouldn't think twice about replacing roof tiles, fitting new doors, completely disassembling their car and such like, the very mention of changing a power socket is enough to send them into a dizzy tremble of fearful anticipation. Although other DIY activities can endanger our health in various ways, we are particularly aware that accidents involving electricity are potentially fatal. The good news is that help is at hand. You need have no more nightmares about how to deal with electricians, what action to take if the power goes or what to do if the lights go off, because this book will show you the way. It explains all you need to know, from understanding the systems in your home, to understanding what the legal requirements are, through to dealing with renegade electricians, replacing circuits and much, much more. Fitting sockets, cables, switches, and fuses, replacing part or complete systems, confidently telling an electrician your requirements and general DIY - it's all possible. Just think... no more electrical worries ever again... exciting isn't it!

ELECTRICITY AT WORK REGULATIONS 1989 (EAW REGULATIONS)

- Regulation 16 COMPETENCE. "All work shall be carried out by competent persons, those with enough technical knowledge and experience relevant to the work being carried out", (summary).
- "Nobody should carry out any work unless they are competent to do so and have had their competence verified or are under qualified supervision", (summary).
- "To comply with the relevant laid down safety standards, (H&S, EAW Regulations) all electrical installations should conform to the latest edition of the IEE Wiring Regulations, BS 7671".
- "On completion and before being connected to the supply all electrical installations MUST be inspected, Tested and Certified to ensure that it complies with the latest edition of the IEE Wiring Regulations".
- "Failure to comply with these Safety Regulations could result in legal proceedings and possible prosecution".
- The office of the Deputy Prime Minister has instigated new legislation "Part P" which comes into force on the 1st January 2005 and places a legal requirement on domestic electrical installers to inspect, test and certify ANY electrical work on completion. This must show compliance with the current Electrical and Building Regulations.

NEW FIXED WIRING COLOUR CODES

All fixed wiring cables purchased after the 1st January 2005 must conform to the new colours below.

- Red live will change to BROWN.
- Black neutral will change to BLUE.
- Green and yellow earth will be UNCHANGED.

This book includes both old and new codes: Red/brown = live and black/blue = neutral.

GENERAL NOTE

A reputable electrician may not feel able to Inspect, Test and Certify any installation installed by a non-qualified electrician and may want to check the installation before carrying out testing.

IMPORTANT SAFETY NOTES

- The isolation of electrical supplies is a recommendation of the EAW Regulations. To make sure that the circuit is safe to work on you must switch off the circuit at the MCB (remove the fuse) and check that the circuit is DEAD before carrying out work. The circuit can be checked by using a safe multimeter set on voltage (certified for use on the voltage) or by plugging in an appliance into the socket that is to be removed. The equipment used to check that the supply is off must be checked on a known working circuit before and after the dead circuit is checked.
- The EAW Regulations state that live working is to be discouraged and only qualified competent electricians should work live and only then when they have put in place all the required safety precautions.
- Replacing consumer units is very dangerous and should not be attempted by unskilled people.

CHAPTER 1

Preliminaries

Your needs

Whether you have just bought a new home or have lived in the same one for years, there will be some electrical work to be done. If you have ever wished for another socket behind the television or above the kitchen worktop, here's how to plan your needs. Every home has different requirements depending on how many people live there, what each room is used for and, of course, the positions of the furniture and appliances. All too often people arrange a room because, for example, the socket for the TV is already in a corner - even if this doesn't work with the furniture, window or lighting. You will also see many homes where extension leads are trailed around the skirting boards or under carpets simply because that was the quickest solution at the time!

Look at the electrics around your house with safety in mind - trailing leads, broken sockets and sparking light switches are dangerous and should be replaced straight away. Walk around your house, garage and garden with a notepad and make a list of all the sockets, lights and leads. Using this, together with the following pointers, you will be able to plan the electrical needs for your home.

Next to each item note any changes you would make if at all possible. Then write down any faults you have noticed or suspect may be a problem, for example trailing wires, burnt sockets, flickering lights or sparking sockets or switches. Note the position of any 3-way plug-in adaptors or bar sockets as these are places where you are definitely lacking wall sockets. Also ask yourself the following questions.

- Do you use power tools or an electric lawn mower outside? If so, is a safety socket (it will be labelled "RCD" for residual current device) fitted? The RCD could save your life if you cut the mower cable by accident or get a power tool wet in the rain, so it's a good idea to have one!
- Do you have a shed or garage in which you would like to have lights or power sockets installed?
- Do you have a fuse or trip switch in the consumer unit (fuse box) that blows regularly?
- Are there dim areas in any of the rooms where you would like to have more light?
- Would you like to have security floodlights on your drive or in your garden, or a lamp post on your lawn?

Go through all the possibilities and write a final list of requirements in order of the most important first. This list can then be priced item by item, so you can work out your budget.

You may be able to do some of the simpler jobs yourself and detailed instructions for doing these are given elsewhere in this book. First find all the references in the book that match your list of requirements. Read each section carefully and decide whether you are competent and up to each task. Remember, there are very strict safety guidelines to follow; if you don't feel competent, don't attempt the task - ELECTRICITY CAN EASILY KILL YOU. Decide how long each task you want to do yourself is likely to take. Then be honest with yourself about whether you really have the time to complete all the tasks you need to in the time you have. Now you will have a plan of tasks to undertake yourself and those to give to a professional electrician.

10 Expert Points

TEN POINTS TO CONSIDER BEFORE YOU START WORK:

1 ELECTRICITY AND KIDS
If you have children, plan your electrics extra carefully as kids will find ways to play with sockets and cables that you've never dreamed of! The lead in a pencil, for example, conducts electricity and can be deadly if poked into a socket.

2 SAFE SOCKETS
Look into sockets with a torch. There should be a small plastic shutter closing off the two lowest holes. The top hole is open for the earth pin to enter. If there is no shutter, then replace the socket.

3 BURNT OUT SOCKETS
Look carefully at the two lower holes in each socket. Do either of them appear to have a brown or black smudge around them? If so, this is burning from overloading and the socket must be replaced.

4 WORN CONTACTS
Plug a lamp into each socket in turn. Wobble the plug a little to see if the light flashes or flickers. If it does, the likelihood is that the contacts in the socket are worn and will soon start to burn. Replace the socket as soon as possible.

5 OLD CONSUMER UNITS
Look at the consumer unit (fuse box). If it is made of wood or brown Bakelite plastic, it could be very old and needs replacing. Ask a professional to check it out and advise you whether it needs replacing or not.

6 NEW CONSUMER UNITS
Do the fuses in the consumer unit have little push buttons or switches? If not, the unit has old-fashioned fuses that have to be replaced when blown, which can be very annoying in the dark when the lighting circuit is out! Consider having a new consumer unit fitted with miniature circuit breakers (MCBs). These are easy to reset after they trip just by pushing them back in.

7 OLD WIRING
Look at the wiring in the loft and at the top of the consumer unit. Is it made of a black rubber like car tyres? This type of wiring will be old now and will have hardened or perished. This is very dangerous and the wiring must be replaced.

8 KITCHEN SOCKETS
In a kitchen, you will need many sockets – sometimes as many as 15 – in order to have power for each appliance. Do you have enough sockets without using adaptors or do you need a few more?

9 SOCKETS FOR VACUUMING
When you are vacuuming the carpets, it is convenient to have sockets in hallways so that the cable need not be trailed around the house. Would you like extra sockets in the hallways?

10 BATHROOM ELECTRICS
In a bathroom or shower room the light switch must have a pull-cord and there must be no ordinary sockets at all. Only an approved-type shaver socket is allowed. Do you need to have sockets removed or the light switch changed?

Restrictions

There are certain areas of the electrical installation in your house that you are not allowed to tamper with in any way, by law. Electrical power may be supplied to your house via underground cables, which pass through the foundations, to the consumer unit. Alternatively it may be supplied via overhead cables from electricity poles. The overhead cables will be fixed to two large ceramic insulators on the roof, chimney or gable end. All the components up to the consumer unit or fuse box belong to the electricity board, which is responsible for their repair. You must never tamper with the cables supplying electricity to the consumer unit or fuse box and you must avoid getting near them when doing other work on the house. If you think the bracket tethering the overhead cables to your house is loose or needs repairing in any way contact the electricity board, who will organize free repair. If you look at the electricity meter and the main fuse you will see that they are sealed with small labels and little lead-stamped seals. It is illegal to cut the seals or labels and doing so could lead to you being prosecuted for theft of electricity. However, the consumer unit or fuse box does belong to you, as do all the other electrical components in the house.

Location

Before you start drilling or chiselling holes in your walls or ceilings you should make sure that there are no cables buried in the same spot. Use a cable tracker, which is a cheap gadget that can be bought from DIY stores, to find where any wiring lies. Do beware though, as it will only give you a guide! Double-check by tapping the wall with a screwdriver handle and listening for a different tone, which can be heard where cable ducting is installed. If your walls are plasterboard on timberwork the wiring will be in the cavity between the plasterboard faces. In this case, double-check the wiring situation by drilling a short (25 mm) hole and peering into the cavity using a torch. As an extra safety measure when you are cutting the plasterboard, use a short 25 mm jigsaw blade so that it's only just going through. If possible you should use battery-powered tools and turn off the main power supply.

Age of the building

Depending on the age of the house you will encounter different problems. In Victorian homes with original horsehair plaster, which is common in most terraced streets, you will find that it is impossible to drill or chisel without big pieces of plaster falling off! Think about surface-mounting switches, sockets and cable ducts. If your house is older than about 1970 it will probably have wooden floorboards. Timber floors are ideal for running cables as it is easy to roll up the carpets, lift a few boards and install new wiring. Don't forget that, if your rooms are wallpapered, installing buried cables and backing boxes will certainly mean redecoration.

Type of structure

Exterior walls are generally built of either bricks or grey blocks and covered with plaster on the inside. In old houses the plaster is white and about 13 mm thick. In newer houses the bricks or blocks are rendered with cement for about 10 mm and then skimmed with 3 mm of pink plaster to give a smooth finish.

Interior walls are often constructed of plasterboard on a timberwork frame. Plasterboard has a cardboard face on both sides, which binds the plaster and makes it possible to cut it with care. If you have tongue and groove panelling there will be a 20 mm gap behind it because it is nailed to horizontal battens. If you look very carefully between the tongue and groove boards you may be able to see the tiny nail heads where the battens run. Cables can be pulled behind plasterboard or tongue and groove panelling with a stiff wire called mousing wire, made of 4 mm nylon with brass loops for tying cable on to the ends.

Older houses have wooden floorboards that are nailed down and easy to lift. In new houses, however, the floors are tongue and grooved chipboard, which is more difficult to lift without wrecking it. Ceilings used to be made of thick white plaster over laths (thin, flat strips of wood), which are nearly impossible to work on without damage – after all they have been there for hundreds of years! Modern ceilings, put up since 1970, are made of plasterboard nailed onto the joists (beams) of the floor above.

Emergencies

Over the years there are bound to be a few occasions when you have an electrical emergency to deal with in your house or workshop – this will generally happen on Christmas Eve or another day when it's impossible to get an electrician out! If possible, it's a good idea to keep a few electrical components and basic tools where they are easily accessible.

EMERGENCY TOOLS AND MATERIALS

- A light switch
- A double socket
- A single socket
- A ceiling rose
- A light pendant and a length of 1.5 mm^2 core cable
- Some fuses
- Screw connection blocks
- Insulation tape
- Earth sleeving
- A torch
- Insulated cross- and straight-head screwdrivers
- Pliers
- Cutters
- Strippers
- Tester screwdriver
- A drill with masonry bits
- A simple multimeter

The sort of repair that might crop up as an emergency might involve a faulty socket that keeps tripping out all the sockets on one floor or a light that does the same to a lighting circuit. If a whole circuit keeps tripping out or blowing a fuse you will need to have a good look around to find out which fitting is causing the problem. Before you start TURN OFF the main power supply and use only insulated tools to investigate the problem. If it's a lighting circuit have a good look at every pendant with a torch. One by one take out each bulb and look at the fitting to see if it is blackened or burnt or if anything is out of place. Next unscrew the pendant and ceiling rose covers and look for loose wires and burnt or broken bits. For socket circuits it is best to unplug all the appliances in the house first as it is likely that one of these is tripping the circuit. Replace the fuse or reset the breaker and see if the power stays on. If it does, then plug your appliances back in, one by one. If one of the appliances makes the circuit trip again, it is causing the fault and you will have to replace it or have it repaired. If all the appliances are alright, then have a close look at all the sockets to check for burning or other damage. Check whether water could have seeped into the electrics, for example, from a leak in the loft or under the floor. Replace the faulty unit.

By far the most common cause of emergency faults is when a cable has been drilled through. A cheap cable tracker can help locate cables before you put in screws and nails, but don't forget they're not very accurate. If disaster strikes as you knock in a nail and your lights go out, do not touch the nail. TURN OFF the main power supply and carefully chisel out a small area around the nail hole until you can get a good view of the problem. Usually the wiring is inside ducting, which can be made of plastic or steel. The ducting will run straight, vertically or horizontally. Trace the damaged cable in both directions to make VERY sure you know where it goes. Each end will probably go to a socket, switch, light or junction box. Use the damaged cable to pull a mousing wire or string through the duct, attach the new cable to the mouse and pull it back. The ends can be rewired into the original fittings. If the cable passes into the loft or under the floor it doesn't need to be followed to its end. Instead, cut the cable in the loft or floor space and fit a junction box. Use a 15-amp junction box for 1.5 mm² cable, a 30-amp box for 2.5 mm² cable and a 40-amp box for 6 mm². You can usually find the cable size by looking at the lettering stamped or printed into the outer insulation, near the manufacturer's name. As a *guide* the following cable sizes are usually used: 1.5 mm² for lighting, 2.5 mm² for sockets, 4 mm² for immersion or storage heaters and showers (10 mm² for some instant heat showers), and 6 mm² for cookers. Look up the section on rewiring the fitting or the junction box (page 80) and make sure you follow all the safety rules. Do NOT do any work unless the main power supply is OFF.

New systems

It must be everyone's nightmare to move into a new home and find that the electrical system is 40 years old. This is a common occurrence so take care when you're looking round a house that an old system has not been disguised by fitting new sockets and switch plates. Have a good look around the fuse box or consumer unit. Signs of an old system are brown Bakelite fuse boxes, push-in rewireable fuses and old black rubber-covered wiring. If you look at the top of the fuse box it is usually possible to see the wiring that goes upwards: if it's black and round instead of flat and grey then beware! Black rubber wiring is a clear sign of an old system because it has not been used for at least 30 years. The problem is that it perishes with age and just crumbles away when disturbed. The black rubber has been replaced by PVC-covered cables. Do not attempt to look behind any fittings, as this will move the wiring, possibly cracking the insulation.

It is not possible to rewire an entire house as a DIY project for the following reasons:

- The wiring circuits need to be planned by a qualified electrician who has a good knowledge of the electrical safety regulations.
- A new consumer unit will probably need to be installed, which means that the electricity board engineer will have to make the new connection between the meter and the consumer unit.
- It is illegal to make this connection yourself because it involves cutting the lead seals off the main fuse or electricity meter.

Remember, the meter and main fuse belong to the electricity board and you must not interfere with them.

Instead, use the Expert Points sections on choosing an electrician and planning a project on pages 15 and 23 of this book and you should be able to get the rewiring done swiftly with all the fittings exactly where you want them. If a complete rewire is undertaken with all the new cables fitted in ducts buried in the walls, it is inevitable that there will be extensive damage to the plasterwork, wallpaper and paint, so allow for the cost of redecorating too. If possible it might be best to stay with relatives or friends in your local area while the work is being done as the mess and disruption will be considerable.

Usually the electrician will bring at least one mate because it is very difficult for one person to install the wiring. It should take two electricians about five working days to rewire a three-bedroom house, not including any redecorating. Remember to have the final quotation and full specification of the quality of the fittings agreed before any work starts. In the specification there should be a diagram of each room showing where you want the sockets, lights and switches. Use coloured stickers around the house to show the electrician exactly where each item on your diagram is; for example, red for sockets, blue for lights, etc. If you do stay in the house while the rewiring is underway you will survive, although you must expect some inconvenience and a lot of clutter and dust.

Part replacement

It may not be necessary to renew the whole system in your house. It is unusual now to find an entire system that has remained unchanged for 40 years. You may just need a new consumer unit, to replace part of the wiring or have a new circuit. These tasks need to be done by a qualified electrician, but before you call someone in and give them carte blanche to rewire your house, do some investigations of your own to find out what is really needed.

For example, the fuse box or consumer unit itself may be old with push-in rewirable fuses, but the wiring may be modern grey or white PVC. In this case it may be possible to simply have the consumer unit replaced with a new model using small trip switches called miniature circuit breakers (MCBs). These are ideal as they are very sensitive, protect the circuit very well and can be reset just by pushing the switch up. The main switch in the consumer unit is the master power switch for the whole installation and can now incorporate a residual current device (RCD). An RCD main power switch is the ultimate in safety devices for your home. An RCD will generally prevent any serious electric shocks but be aware that children, old people and people with heart problems are susceptible to smaller levels of current.

In other cases it may be that some circuits in the house have been replaced with PVC wiring while others are the old rubber-covered cables. If this is the situation you can have the old circuits replaced and leave the PVC cables as they are.

If your house has a fairly new system but you require more sockets and lights because there is only one socket in each room or you build an extension, you may need to fit a new consumer unit that has more circuits. However, if the existing unit has some spare slots these can be used for the new circuits.

If you do need a new consumer unit you will need certain information to specify what you want. Count the number of fuses or breakers in your consumer unit and then add one for every extra circuit you require. For example you might require an extra socket ring main, lighting ring main, instant shower or wall heater. Each circuit needs an MCB which fits into a slot in the consumer unit, so, for example, if you need a total of 12 circuits you need 12 MCBs and you must specify a 12-way consumer unit. You will also need to specify the trip rating for the main power switch on the consumer unit. It will usually be either 80 or 100 amps, depending on the size of the electricity board's fuse (the one with the lead seal on it) and the cable between the meter and the consumer unit. You must check the fuse size, which is printed on the casing. The trip rating of the main power switch for the new consumer unit must not be greater than the fuse size, because otherwise the main electricity board sealed fuse could overload. Once you have all this information you can specify the consumer unit you need, as for example, "a 12-way consumer unit with an 80-amp RCD main breaker".

Please read the section on "New systems" (pages 10–11), as many of the procedures are the same for just replacing a circuit or part of a system.

Fitting major items

There will probably be a time when you want to add a major item to your electrical system. Typical projects might be installing an instant shower (see page 110) or outdoor floodlights, or taking power to a workshop or garage.

A new cable run will also be needed, so plan the route from the consumer unit to the new installation. Remember that it is usually advisable to bury cables in ducts and that this will mean cutting into the walls and floors or ceilings. It will inevitably make a mess. It is possible to fit the cables in surface-mounted ducts but they can make the place look a bit like a factory, so do consider this carefully.

If you want to run power to a workshop or garage you need to make a list of the equipment you plan to plug into this new system. Note down all the equipment you want to connect and find out how many watts (W) of power each one consumes. Remember 1 kilowatt (kW) is 1000 W. You will never use all the equipment at the same time, but you should have a think about which items you are likely to want to use in combination, for example you might want to use a heater, lathe, dust extractor and lights together. Add up the watts that these units will draw and divide the total by 230. This gives you the smallest size, in amps, of the supply cable and the fuse or MCB in your main consumer unit that you need.

It is usually possible to fit a 30- or 40-amp fuse or MCB in the main consumer unit and run a 4 or 6 mm^2 twin-core and earth cable to the smaller consumer unit in the garage or workshop. The cable should be in ducting or attached to a tensioned wire rope between the buildings. If you go for the overhead method it must be high enough so that it cannot be damaged by children, cars, garden tools, etc. Cable for running overhead should be at the very least an "SY" type, which has a wire-armoured braid under a clear PVC covering. Alternatively, the cable can be buried underground, but it should be at least 1 m deep and it should be fully wire armoured.

The electrical system from the house will not support the use of three-phase, motor-driven equipment or any requiring more than 80 amps, such as a large welding machine. The supply to your house is single-phase 230 volts (V) ± 10%, whereas most large workshop machines such as lathes, milling machines and welders operate on 3-phase 415 V. If you want to run such equipment, you will need to contact the electricity board and discuss installing a totally new supply to the workshop or garage. Inconveniently, three-phase power is not available everywhere and the electricity board may have to run new cables from a pole, junction box or substation. Needless to say, this will be very expensive!

CHAPTER 2

Plans

At some point you will need the services of an electrician, either to check your work or to undertake the jobs that you cannot do yourself. Electricians are the same as any other tradesmen - there are good ones and bad ones! You may have seen television "fly on the wall" documentaries filming dodgy electricians at work, bodging jobs and charging ridiculous prices. These people do exist and will blatantly rip you off. However, worse still is the incompetent electrician who is not working to the proper standards.

Electricians have to take examinations to ensure their work and knowledge is up to standard, but it can be difficult to check these qualifications. A properly qualified electrician will not take offence if you ask to see a copy of the certificates and will be proud to show them off. "Mr Dodgy" the electrician will bluster and make excuses as to why he can't produce the certificates - use someone else. It is also a good idea to ask people you trust if they can recommend a good electrician.

There are some jobs that you can do yourself if you are skilled at DIY and have a good understanding of the task you need to do. If you fully understand the wiring, the type of installation and what you need to change or modify, you can then plan the job. To check whether you can do a small job, **switch off the circuit at the MCB (remove the fuse) and check that the circuit is DEAD** before carrying out any work or removing any electrical components. As you do so, make a drawing of the existing wiring using coloured felt-tip pens and label all the connections, as they are written on the fitting or appliance. You will also need to consider the parts, tools and techniques to be used. Make each fitting safe again by closing it up and turn the power back on. Your drawing will be useful for two reasons - it can be used for planning the new circuit and also for retracing your steps if the job cannot be completed for any reason. It is not practical to have the power off for long lengths of time, so the option of retracing your steps should be kept in mind just in case.

A few examples of possible DIY tasks are fitting new sockets, switches, ceiling roses, pull-cords and outside lights. Jobs that you should not attempt yourself are replacing the consumer unit, rewiring the house, etc. You must never undertake any electrical task if you are not 100% sure about how to proceed.

Whatever work you undertake yourself, it should be checked by a reputable electrician. You will obviously have to pay for the inspection, testing and written certification, so write down any faults that the electrician finds so that these can be corrected. Before selecting the electrician, refer to the Expert Points box opposite.

Requesting plans and details

As with discussing and agreeing a job with any tradesman, you must be clear about what you want the

10 Expert Points

TEN POINTS ON FINDING A GOOD ELECTRICIAN:

1 QUALIFICATIONS
Ask to see the electrician's IEE Wiring Regulations, BS 7671 certificate and C&G 2381 certificate of qualification as this will confirm that they were properly trained.

2 RECOMMENDATIONS
Make sure that at least two people you know have used the electrician and were happy with the jobs done.

3 PROFESSIONAL APPEARANCE
Have a look at the electrician's transport. Is it a nice clean van with a trading name on the side or a rusty estate car? Electrical work should be cleanly, precisely and safely done and the transport will reflect the electrician's standards.

4 CONFIDENT SKILLS
When you ask questions does the electrician hesitate and bluster or have confident answers?

5 CONSIDERED QUOTES
Beware of electricians who look at the work for two minutes and then give an "off the cuff" quote. This doesn't indicate proper planning – it is next to impossible to price work without measuring cable runs and checking price catalogues.

6 CONTACT DETAILS
Do not use an electrician who will only give a mobile phone number and not a full address. You will need to be able to contact the electrician if something goes wrong within the guarantee period. Most fittings are guaranteed for at least one year.

7 SCHEMATIC PLANS
Ask for a schematic plan from the electrician and check it matches your requirements before work begins. The plan will show each cable run and individual fittings and appliances with their connections to the existing consumer unit or electrical system. It will not show every core of wiring.

8 REGISTRATION NUMBER
Most electricians are members of the Electrical Contractors' Association (ECA) or the National Inspection Council for Electrical Installation Contracting (NICEIC). If you ask for the electrician's registration number, you can check them out on the Internet.

9 INSPECTION CERTIFICATE
The electrician should be willing to give you a signed inspection certificate bearing their registration number once the work is complete. This certificate will show the results of a number of tests for electrical safety and should be kept safely as you may have to produce it if you sell the house.

10 WORTH THEIR WEIGHT IN GOLD
The electrician should be someone you feel comfortable with and who clearly listens to your requirements. Good tradesmen are truly "like gold dust", so ask around your friends and neighbours to seek out the person you need.

electrician to do and how you want the work done.A good electrician will advise you on what is and is not possible and whether the changes will have any wider implications for the whole system.Ask the electrician to explain all the aspects of the project in detail so you can consider possible implications and check out the options at your leisure.

All electrical components can be bought in different grades and qualities. As with most things in life, you get what you pay for with electrical equipment! If you can afford to, it is better to go with good brand names and pay a little more - especially for light switches and sockets. Explain to the electrician that you would like good quality fittings with brand names like MK or GE.

Discuss your requirements as to whether cabling and fittings should be on the surface or buried in the wall. Also specify if you would like the plastering and decorating to be finished "as found". To save money you could make good the decorating yourself, although this is a messy and fiddly job! If wallpaper is damaged, it may not be possible to find a match - so be prepared to repaper the room.

Ask the electrician for a complete and detailed quote, including a start and finish date.

Quotes and estimates

Care should be taken when getting a price from an electrician. With luck, recommendations from friends or relatives will have given you a few possible candidates to take on the work who can be trusted to do a reasonably priced, good quality job.

Unfortunately, there are some tradesmen who will try to use the difference between a quote and an estimate to their advantage. This practice is easy to miss and will invariably result in things costing a little, or a lot, more than you thought. A quote is a fixed price; it is the same as going into a shop and asking for a new bed. You are given a quote and this is the price you will pay. An estimate is different because it is essentially just an educated guess and is not necessarily the figure you will finally have to pay. Traditionally, the final price you will have to pay will be within 10% of the estimated price - that is to say you could have a bill that ends up being 10% cheaper than the estimate. However, human nature being what it is, you could equally well find that the final bill is 10% more than the estimate.

When you have selected three electricians you would like to consider for the job, you will need to give each of them your specification and ask them for a "quote in writing, to complete the work in the attached specification to the proper standard". You should also specify that you will require a "Part P" test certificate. To summarize, a quote is a fixed price given against your specification, it must be given in writing and should refer to the specification stapled to the back of the quote.

You must take care when producing your specification. Read the relevant sections on these pages and page 23 thoroughly and make sure you have included everything you require. Go round the house with the electrician and discuss the work required in situ before they give you a quote. Ask the electrician to read the specification and tell you if anything is inaccurately expressed, cannot be done or can be done in a better way. If their reasoning is sound, amend the specification accordingly and send them a new copy as the basis for the quote. This way the electrician knows what is required and exactly how you expect the work to be finished.

The quote should include VAT only if the electrician is VAT-registered. Get the VAT number from the letter head and check it out on the Internet at www.hmce.gov.uk. It is a common scam to give a bogus VAT number and charge the VAT but not pass this on to Customs and Excise. It is not possible to get back VAT money that has been falsely claimed without an order from Customs and Excise, so if you do pay it, you could lose it altogether.

Ask for a test certificate for the property after the work has been completed. This should also be put somewhere safe for when you sell the house. In Scotland it is necessary to produce a buyer's pack for the intending purchaser, containing various structural survey reports as well as test certificates for damp prevention and electrical installations. This may become the law in England too, so keep the certificate safe because it may save you money.

It is normal to pay anything up to 50% of the quotation price before work starts to enable the electrician to begin buying the cabling and other necessary components. The balance should be paid when the job is finished and you have been given the test certificate. Always make sure the terms in the specification are to your complete satisfaction.

Although a firm fixed quote is the ideal, many electricians will only be willing to provide a very rough and ready verbal estimate. They argue that there are usually so many unknown variables, such as poor plaster and bad wiring hidden away in the wall cavities, which might push up the costs, that they really cannot give a fixed price. If all three of your chosen electricians veer away from giving a fixed written quote, consider a few relevant factors. If they have been recommended and work locally, they will have a reputation to keep. Their reasons for not giving a written quote may be genuine or they may simply be too busy for paperwork. If you want to book time in their busy schedule, you might have to consider accepting an estimate after all. Of course, it is not always the case that an electrician is going to rip you off - sometimes they simply hate paperwork and prefer to shake on a deal, but you do have to be on your guard. If you have no other choice than to accept an estimate, then at the very least make sure that you get a written agreement, setting out in clear terms the maximum price that you will have to pay. So, while the ideal situation is to have a written quotation provided, if the electrician will only give you an estimate, then it must be backed up with a written maximum price. If the electrician will not agree to this, then you can rightly say "no thank you" and look for another electrician. A charming manner or a smart suit are all good signs, but not as good as a signed and dated written quotation.

Standards of work

The finished work must be safe and fit for purpose. It must conform to certain regulations and codes of practice as set out by such bodies as the Institute of Electrical Engineers (IEE) and the Electrical Contractors Association (ECA). Although there is no legal code that requires your chosen electrician to belong to such a body, they should nevertheless provide you with a certificate of minor works, British Standard 7671. This is a guarantee to say that the job is up to standard. Ask for this document just before you hand over the money.

Be mindful that you only hold power while you hold the money. The time to start asking questions about the standard of workmanship is not when the job is done, but the very moment that you see a problem. So, for example, if you see that a socket is cracked or crooked, then that is the time to start asking questions. There is no point letting the job chug through to completion if you are in any way unhappy about the standard of work. The best way forward is not to spend all day hovering over the electrician, but rather to make notes after the working day and then to ask questions in the next convenient rest period. A potential problem, of course, is that while you can see with your own eyes if a switch is crooked, a socket is blackened and sending out sparks or the lights are flickering on and off, there is no way that, as a lay person, you could tell if the cables are up to standard or the earth connections are sound. If any of the components that you can see are crooked, scratched and generally less than beautiful, then the likelihood is that parts of the work that are hidden from view will also be less than satisfactory.

Remember, the key words are "safe and fit for purpose". So, in order to recognize what is and is not an acceptable standard of work, read through the relevant sections of this book and you will have all the information that you require.

How to avoid cowboys

Long, lean and lonesome, these troublesome electrical cowboys are everywhere and yet at the same time they are difficult to identify. The problem is that while they look like electricians and sound like electricians - and some even genuinely think that they are electricians - they can't easily be spotted. And, of course, while we all use the term "cowboy" in a jokey sort of way, the thing to bear in mind is that we are not talking here about the honest and friendly type of goodies that we love - these characters are most definitely the mean baddies that we hate.

To make matters worse, there are two types of electrician baddy. There is the well-meaning but under-qualified guy who thinks that he knows what he is doing, but doesn't, and there is the guy who doesn't give a damn. The first may give you a low quote and unintentionally do a bad job, while the second will knowingly do a bad job, swiftly take your money and move on to new pastures and the next gullible customer. While we might feel sorry for one and loathe the other, they are both dangerous. The following pointers will help you avoid both of them:

- Never employ the person who arrives at your door uninvited and points out all your electrical problems.
- Avoid the person who wants large amounts of cash up front.
- Stay away from anyone who gives you a mobile telephone number but not an address.
- Be careful about anyone who is offering to do a swift job for cash.
- Avoid those who are only prepared to work out a rough and ready estimate on the back of an envelope.
- Try phoning the firm during working hours to see if they are well organized and able to take messages.

Who are you letting into your home?

While there are many kind and helpful people out there, there are also dishonest people who will try to relieve you of your money. Always ask for identification, including a name, address, and photo-identity card before you let strangers into your home. If you live on your own, then make sure that you are not alone when the electrician makes their first call. If you are worried in any way, then ask the local council to supply you with names and addresses of suitable electricians. Councils are not allowed to recommend individuals, but they will usually provide you with a list of tried and trusted electricians.

The good news is that while the advice here may suggest that you will be battling against all manner of robbers, the truth is that the majority of electricians are just trying to do their best to make an honest living.

Finance

Cash might be the easiest way for you to make a payment and is what every small business person is looking for. It is easy and convenient if the job is small or you are buying items to do the job yourself. However, the best way to pay for large expensive jobs is through the banking system. You hand over a cheque or card and the electrician gives you a written, dated and signed receipt. If a friend or family member is on hand to witness the transaction, then so much the better.

While most of us have savings to cover small to medium jobs, there may be times when you need to borrow the finances. If the job is large, then you have the choice of extending your mortgage or getting another type of loan secured on the property, or of arranging an unsecured loan that you will pay off in the medium term.

Grants

To a greater or lesser extent, local authority grants are also available. If you have a priority need because you are very elderly or housebound, have very young family members or have a disability, priority funds and emergency grants are available. The local authority has a duty of care to make sure that the elderly and vulnerable are safe in their homes.

Grants are available to the elderly who have electrical systems in need of repair or replacement, but little or no independent income. They are also available for those with a disability.

If you need special lighting or heating or special electrics in the bathroom because of age or disability, you should contact your local planning authority to discuss your requirements, explaining your age and situation, and ask them to make a visit to assess your needs. The secret that guarantees swift action is to make sure that you get full names of the contacts you make. If you can get the name of the head of the appropriate department, then so much the better. An initial phone call will establish a contact name and address, but then send a letter to the head of department setting out what you consider to be the problem. Keep copies of the letters. If you want to make absolutely sure that wheels start turning, pay for a postal service that guarantees signature on delivery. A representative from the local planning office will call and arrange for a qualified electrician to look at the electrics in your house. A written report will be sent to the planning authority. If the electrics are in any way dangerous, then the planning authority will authorize the work to start immediately.

If you qualify for them, grants are yours by right - so don't be put off or intimidated by red tape.

Payments

If the job is a small one worth no more than about £100 then payment isn't really much of a concern. You come to an agreement, you see with your own eyes that the task has been carried out successfully and to the required standard, and then you hand over the money.

There might be more cause for concern when the price for the job runs into hundreds or perhaps thousands of pounds. The best way forward in this case is, once you have agreed on the general terms of the agreement, to write an extra clause into the contract to cover the way in which the money is to be paid. For example, both you and the electrician might prefer staged payments; you might pay one quarter of the total when the materials have been delivered to your address, one quarter when the cables have been put in place, and the balance of the money when the installation is running and you have been given the test certificate, probably a few days after the work is complete. If the electrician finishes the job and the system fails but the final stage payment is yet to be made, you must then state in writing how much money you are going to withhold from the final payment and the reasons for holding it back.

Day work payments

Just occasionally an electrician might suggest that you pay him an hourly rate. While this is a good option for a small swift task such as changing a socket or switch, it is not such a good option if the work involves jobs like chopping through plaster or crawling through a loft, for which the time needed might be less easy to quantify. If there is a chance that a job could get complicated, then avoid an hourly rate. However, if the task is simple and straightforward, such as having all your single sockets changed to double ones, an hourly rate may be quite a good option for two reasons. You will know precisely how you stand as regards the cost of materials and all labour costs will be covered in the agreed hourly rate. There won't be any hidden costs and, if need be, you can stop the work whenever you like.

Insurance

As a homeowner you are legally responsible to some extent for the wellbeing of your guests, neighbours, those entering your property and passers-by, as well as for the condition of the actual fabric of the building.

All responsible, qualified electricians must be insured against injury, building damage and suchlike. And so, if you invite an electrician into your house to do work that involves such things as crawling through the loft they need to be insured. It is important to know exactly what their insurance covers, so make sure that the insurance details are set out on the contract. You should consider what would happen if the electrician should fall off a ladder or through the ceiling, injure one of your guests or even burn the house down... It is best to phone your own building insurance company too to find out how you stand. If you have any doubts, then take out extra insurance and make sure this is in place for the duration of the whole project.

Health and safety

Electricity is potentially very dangerous. Get it wrong and it can give you a shock, cause a fire or even kill you. For the short time that the electrician is in your home, your home will become a dangerous place. Danger is normally present in the form of ladders that can be tripped over, live wires that have been left unprotected, loft hatches that have been left open, tools falling and so on. In addition, some of the materials and procedures used during electrical work are dangerous. There will be heat, dust, fumes and poor ventilation to contend with. While the electrician has a duty to leave your home in a safe condition at the end of each day that he has worked there, your best course of action is to assume that everything is a potential danger. If you see things like a bare wire, open tins of chemicals, or power tools plugged in, treat them with all due caution.

Snagging

Snagging is the term given to the procedures that are undertaken to correct any errors that may have arisen, thus bringing the agreed job to a satisfactory conclusion. Snagging is definitely not about adding extra tasks to the work or changes to the specification after the contract has been agreed. You can't ask the electrician to fit brass switch plates when you have both agreed on plastic ones; or at least you can ask to have them changed, but you can't rightly expect that to be done without incurring extra costs.

While on a small job snagging might involve no more than asking the electrician to straighten up a single light fitting, on a large job it might involve making dozens of small adjustments. A good way forward is to make a list, with one copy for the electrician and one for your own records. Let the electrician know at the outset that you are keeping a snagging list. Then, if you spot, for example, a number of crooked sockets, some badly fitted covers, fittings that should really have been better placed, pendant lights that are obviously too low, a chipped switch or a light that doesn't always work when you switch it on, you have a couple of choices. You can either ask the electrician to make good as the work proceeds or you can get him to bring the work into good order just before you make the final payment. Outstanding items on a snagging list are a common reason for the withholding of the final payment.

Trade agreements and disputes

While the average home electrics job can be done for a relatively modest amount of money, some jobs do run into many hundreds or thousands of pounds. In these situations it's a good idea to make sure that your chosen electrician belongs to the appropriate professional bodies and quality mark schemes. Ask if the electrician belongs to such a body and check out the details just to make sure. Although such affiliations won't help you if the electrician is criminally negligent, to a limited extent they will give you confidence that your chosen electrician is a responsible individual who is going to do his best and work to the required standards.

DIY options

DIY is not everyone's cup of tea! It is very easy for people who can fix most things and have a fortune in snazzy tools to talk about how to do this and that as if it's all easy. While one person can put in a socket ring main, others struggle to change a plug and that's the way life is!

Electrical work is a little more logical than putting up a shelf and if you've ever put a flatpack chest of drawers together from instructions in a rather obscure version of English you should be fine. The secret to doing electrical work, as with all DIY, is to have a long, careful look at what the job entails with safety in mind. Don't be tempted to rush in and start pulling things apart! Make a few diagrams with coloured pens and work out what is going on as things stand. Use this book and your diagrams to plan your work and make sure that you feel confident about what you are going to do. **If you are unsure in any way, then use this book to write a specification and engage a properly-qualified electrician to do the work for you.**

10 Expert Points

TEN POINTS TO HELP YOU PLAN YOUR WORK:

1 TURN OFF THE POWER FIRST
NEVER EVER work on the system when it is live. Electric shocks do really hurt! While the shock itself probably won't kill you, falling off the stepladder in surprise and hitting your head on the corner of the cooker probably will finish you off! Always TURN OFF the main power supply first.

2 UNDERSTANDING CIRCUITRY
Do you understand the part of the system you are working on? Electrical circuits are easy to understand when they are drawn out. In reality all you could see is a jumble of red/brown, black/blue and green+yellow wires in a box. Don't let this confuse you. Work out where each wire goes and draw up a circuit diagram yourself.

3 THE BEST TIME TO START
Do not start electrical jobs at 4.30 on a Sunday afternoon. There won't be many hours of daylight left and the shops will be shut, so getting that last minute item you forgot will be out of the question. Plan the work and have everything ready for an 8 o'clock start on a Saturday morning or a weekday in your holidays.

4 RETRACING YOUR STEPS
Do you have a plan to undo what you've done if necessary? Make a diagram of the original system so that you can always put things back as they were if you hit a snag.

5 USE GOOD TOOLS
When buying tools for electrical work, buy good ones. They do a good job and the work will be all the easier for it. Remember – "buy rubbish – buy twice".

6 FINDING ELECTRICAL FAULTS
A simple digital multimeter (sometimes referred to as a continuity tester) is the secret to electrical fault finding. Buy one and make sure you know how to use it.

7 PEACE AND QUIET
Plan the work so you have peace and quiet; send the kids out for the day and put the dog in the garden. However, do make sure another adult is within earshot. NEVER work on electrics alone in case something untoward happens.

8 STAGING THE WORK
Prepare as much of the work as you can before you turn the power off. For example, roll back the carpets and lift floorboards, cut into the walls and fit all the ducts, and run the mousing strings through the ducts in advance.

9 BUYING IN BULK
If you think that your house is likely to need a few jobs over the years, buy a whole 50-metre roll of both 1.5 mm^2 and 2.5 mm^2 twin-core and earth cable from a wholesaler. It is vastly cheaper this way and you can store the spare cable in the loft in readiness for the next job.

10 THINK OF THE FAMILY
Remember, your family will want to watch television, have hot water and the computer on every evening! Plan your DIY carefully, prepare what you can, get the job done properly and the power back on swiftly or you'll be in the dog-house!

CHAPTER 3

System options

What's going on in your house?

In the average house there are usually three basic electrical systems - the power circuit, the lighting circuit (or circuits) and the dedicated circuits.

If you have a good long look at your consumer unit or fuse box, which might be under the stairs, in a cupboard high on the wall in the hall, in a special box on an outside wall or in the garage, you will see that there are two or three cables running in at the bottom of the box and groups of cables running out at the top. If you study the top cables you will see that they range in size from 1.00 mm^2 to 10 mm^2. Cables are specified in millimetres squared; this is the cross sectional area of the copper in each core and is marked on the outer sheath. A larger cross section is used for heavier circuits. If your house has been correctly wired, the size of the cables will, to a great extent, tell you about the circuits in your electrical systems. If you know which type of cable is used for the different types of circuits (see the "Cables and circuits" box below), you can be fairly sure that if, for example, you can see a 2.5 mm^2 cable, you are looking at a ring circuit system rather than a radial circuit system (see "Power: ring circuits" on page 25 and "Power: radial circuits" on page 28). At the very basic level, the size of cable should give you a reasonable idea of what is going on electrically in your house.

Cables and circuits

If you are a raw beginner, it's a good idea to label as many of the cables in the consumer unit or fuse box as possible. By careful observation you can work out exactly how your home is wired. You can test out the purpose of each circuit by turning OFF the main power supply and then pulling out a fuse or turning off the MCB switch. Then go and test the lighting, sockets or appliances that you think should be on that circuit. If they don't work, you have correctly identified the circuit. If they do, you must look elsewhere. And so you can continue, working from one side of the consumer unit or fuse box and running through the sequence of procedures as just described, until you have identified all the circuits with their respective MCBs or fuses and cables.

CABLES AND CIRCUITS

- **1.00 mm^2 cable** is used for lighting circuits
- **2.5 mm^2 cable** is used for ring circuits, storage heaters and immersion heaters
- **4 mm^2 cable** is used for radial circuits
- **6 mm^2 cable** is dedicated to circuits running to small cookers and small showers
- **10 mm^2 cable** is dedicated to circuits running to large cookers and large showers

The next step in this voyage of discovery is to have a look at what is going on behind the sockets and ceiling roses. Again switch off the circuit at the MCB (remove the fuse) and check that the circuit is DEAD. Use a small electrical screwdriver to remove one of the socket covers. Look at the number of cables entering the mounting box. If there are two cables, then you have a ring or radial circuit. If there is one cable, then you have a spur circuit or the end of a radial. And much the same goes for the ceiling roses. One cable running to a rose tells you that you have a junction box lighting circuit, while two or three cables tell you that you have a loop-in lighting circuit. By inspecting all the sockets and ceiling roses in turn, you will be able to build up a picture of the types of power and lighting circuits, and how many lighting circuits you are dealing with.

Of course, if at the end of this exploration you see that there are anomalies such as lots of sockets strung out on a spur from a single socket, power sockets coming off the lighting circuit or the wrong size of cable running to the cooker, then you can be pretty sure that someone has made a bad job of the electrics. Hopefully, it won't be as bad as this real life example: a flat in which a gleaming new cable ran up the wall from a cooker to the ceiling - just the right size cable and beautifully fitted in trunking - but unfortunately, once inside the loft, fixed with insulation tape to a much smaller cable that snaked out through a hole in a party wall to the house next door!

Power: ring circuits

A ring circuit is the modern method of wiring strings of sockets to distribute power around a building. Technically, a ring circuit is a parallel system with power fed from both ends of the wiring at the same time. This has a number of distinct advantages over spur and radial wiring, the main one being that smaller cables can be used. Smaller cables are much easier to route through ducts as they bend into tighter corners and more cables can be fitted in each duct. Copper wiring is expensive so using a smaller size cable is also much cheaper.

A ring circuit can be identified by looking in the consumer unit or fuse box. Turn OFF the main power supply first, of course! If you can see two red/brown cores from cable of equal size connected to the MCB or fuse, you have found a ring circuit. The live, neutral and earth wires leave the consumer unit, travel round all the sockets in the ring and come back to the consumer unit or fuse box again. In this way, power is fed into both ends of the ring. The effect is that, for example, 2.5 mm^2 cable on a ring circuit will carry the same current as a cable twice the size on a spur.

It can be difficult to identify a ring circuit from looking in the fuse box as it could also be a radial circuit. The radial circuit's "bicycle wheel" design has the main fuse (at the centre of the wheel) with two or more red/brown cores under it, each one going out like the spokes of the wheel and feeding some power sockets. This type uses more cable than the modern ring circuit.

10 Expert Points

REMEMBER THESE POINTS WHEN TESTING CONTINUITY AND REPAIRING BREAKS:

1 LIVE WORKING
It is DANGEROUS to work on LIVE circuits. The EAW regulations state that live working is to be discouraged. Only qualified, competent electricians should work live and only then when they have put in place all the required safety precautions.

2 CONTINUITY TESTING
To test a ring circuit for continuity, switch off the circuit at the MCB (remove the fuse) and check that the circuit is DEAD. Use a multimeter set for "continuity test". Take the pair of red/brown cores from under the MCB or fuse terminal and connect one to each of the multimeter probes. If the circuit is continuous, the meter will read "00.0", and will buzz if it has a buzz facility. Repeat this test with the pairs of cores for both the neutral and the earth, which should also be continuous. The resistance of the wiring is measured in Ohms. The expected readings for c.p.c (earth) continuity are very low Ohms and for an open circuit very high Ohms.

3 CIRCUIT BREAKS
If the ring circuit has a break, the cause needs to be found straightaway. A break in a ring circuit can cause overheating and overload the wiring. Get the problem fixed immediately.

4 BREAKS IN POWER CIRCUITS
A break in a ring circuit is usually to be located in one of the sockets. Turn OFF the main power supply. Check each socket on the ring in turn. Remove the socket cover. Give the wires a good pull and tighten any loose terminals. Check the circuit for continuity before turning on the power.

5 MULTIPLE BREAKS
Remember that there may be more than one break in a circuit. Just because you have found one fault doesn't mean there can't be another!

6 THE USUAL PLACE FOR BREAKS
A break will usually occur in a wire right at the end of the insulation where wire strippers have nicked the copper slightly and made a small weakness.

7 SECURE CONNECTIONS
One way to improve the holding power of the terminal screws in a socket is to strip the wires to 30 mm and fold the copper in half with pliers. This will give the screw twice as much to grip.

8 CONNECTING WIRES TO SHARE A TERMINAL
When there are two or more wires that need connecting to one terminal screw, strip the ends to 20 mm and twist them all together with a pair of pliers to make a sort of barley twist. This will help the terminal screw get a good grip on the copper cores.

9 WATCH OUT FOR RADIAL CIRCUITS
Take care that the circuit you are checking for continuity is not actually a radial circuit with two legs. This is just a couple of sockets or lights wired into one fuse. It will not be possible to get a continuity check on either the live, neutral or earth pairs in a radial circuit.

10 CHECKING IT'S A RING CIRCUIT
If the continuity test proves negative on the three cores, you may wonder if you are really dealing with a ring circuit. Inspect all the sockets. If there are at least two lives, two neutrals and two earths connected to each terminal, then it is a ring circuit. If there is, in any of the sockets, only one wire per terminal it is the end of a spur or a radial circuit.

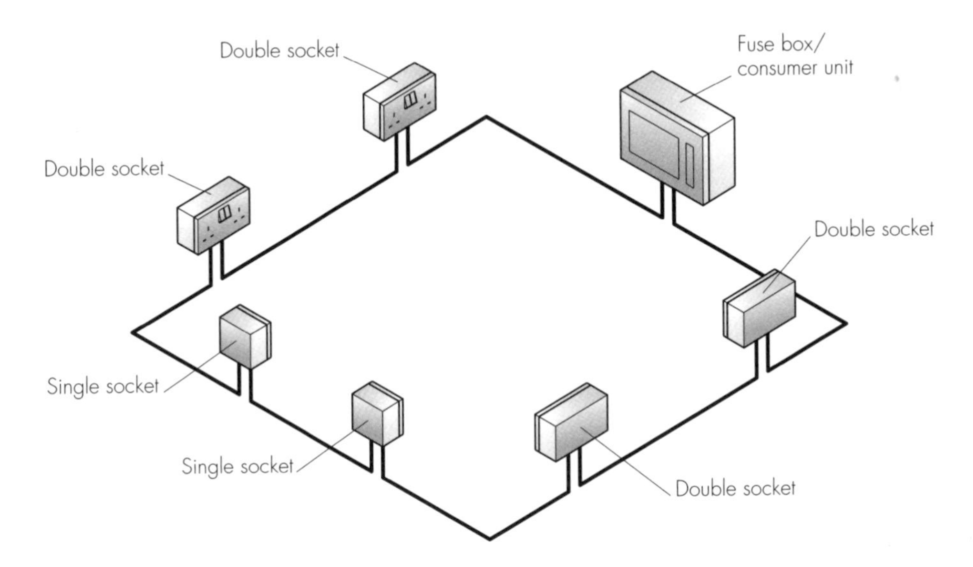

A simple ring circuit.

Study the diagram of a simple ring circuit (above). If you follow the red/brown live wire, you can see that it travels from the MCB or fuse in the consumer unit or fuse box to the first socket. At the first socket this live wire is under the live terminal, which is marked "L", along with the live wire going to the next socket in the ring. The copper ends should be twisted tight together or both be folded double with pliers before putting them under the terminal screw. Read the section on "Connecting cable to sockets" on page 70 for details on how to make the connections. This live wire can then be traced around the entire ring of sockets, connected at each live terminal. From the last socket in the ring, the live wire returns to the same MCB or fuse it started from. This is why you will always see two wires under the MCB or fuse on a ring circuit.

The black/blue neutral wire and the green+yellow earth wire each start on a terminal rail in the consumer unit or fuse box. There is a terminal rail for neutral and one for earth. They look very similar and are usually a strip of brass or copper with many screw holes drilled along the length. You will see that all the neutral wires are connected to the neutral terminal rail and all the earth wires are connected to the earth terminal rail. The neutral and earth wires start in the consumer unit or fuse box and, just as the live wire did, travel from socket to socket and eventually back to the consumer unit or fuse box again.

It is extremely dangerous to have a break in the ring because it can overload the wiring. Refer to the Expert Points panel on the opposite page to check how to identify and remedy a break in continuity.

Power: radial circuits

A radial circuit is an old-fashioned method of wiring up a house to distribute power to the sockets. In this type of circuit, the cable starts at the fuse box or consumer unit, supplies power to a number of sockets along its route and then ends at the last socket on that route. The average small house might well have had two such radial circuits - one radiating out for the upstairs sockets and another for the downstairs sockets.

Even though the radial circuit is an old-fashioned method and your home may have been wired up with a ring circuit, if you want to run a new circuit out to a garage, workshop or other outbuilding and there is space in the consumer unit or fuse box, then a new radial circuit is a good idea. The best option is to use a 20-amp MCB or fuse with an RCD and to wire up using 4 mm² cable. The new cable will run out from the consumer unit or fuse box, with the circuit being protected by the 20-amp MCB or fuse, to feed three or four sockets. The cable will end at the last socket along the line.

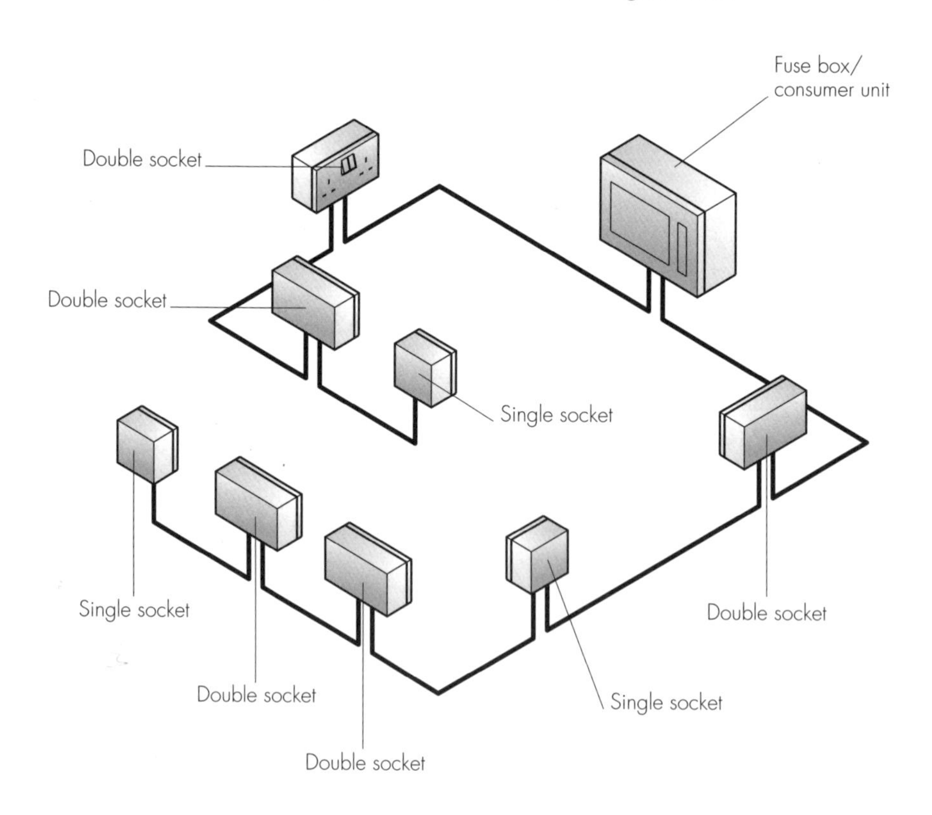

A radial circuit.

10 Expert Points

IMPORTANT THINGS TO REMEMBER ABOUT RADIAL CIRCUITS:

1 TAKING PRECAUTIONS
If you are worried about your wiring and plan to investigate, then tell a friend or family member what you are doing, make sure that you have a good torch to hand and turn OFF the main power supply.

2 SAFETY IN SILENCE
A good tip is to plug in the TV and radio at various points around the house and turn up the volume before you turn the power off at the mains. The silence will serve to remind you that you have turned off the power.

3 LIMITATIONS OF RADIAL CIRCUITS
Although a radial circuit wired in with 2.5 mm^2 cable from a 20-amp MCB or fuse can supply a good number of sockets, it must not run across an area greater than about 20 m^2.

4 REPLACE OLD CIRCUITS
If you live in an older house and the power is distributed on radial circuits, then the chances are that the wiring is approaching the end of its life and needs replacing with ring circuits. If you find that you do indeed have a radial circuit, with rubbery looking cables set in metal tubes and brown plastic switches and sockets, then it needs replacing.

5 GETTING IN THE PROFESSIONAL
While it may be a good idea to make changes to an old radial circuit, this is probably one of those occasions when you need to bring in a professional electrician.

6 LEAVE ALONE OR REPLACE
If the radial circuits are very old, the cables are very likely to be brittle and/or cracked and the rubber sheaths will probably be disintegrating.

7 STRIPPING OUT OLD CABLES
If you have asked an electrician to replace the radial system with a ring circuit, then you can cut your costs by offering to remove the old wiring. The best course of action is for the electrician to cut off the old wires and to clearly label the ends, so that you can quietly spend time stripping out the old cables.

8 RCD PROTECTION
If you have decided to run a new radial circuit out to a garage or workshop, then make sure that you fit an in-line RCD to protect the circuit with a switch fuse unit.

9 LONG RADIAL CIRCUITS
If you think that a new radial circuit to a garage or workshop is going to get a lot of use and/or extends a long way from the house, then use a 4 mm^2 cable.

10 USING RADIAL CIRCUITS FOR LIGHTING
If you decide that you need a new radial circuit to a garage or workshop to feed both the sockets and the lights, you can run a lighting spur off the socket supply.

Power: fused spur

A fused spur is a way of tapping into a ring circuit for a permanently wired-in appliance. Typical items that are wired into a fused spur are cooker hoods, doorbell transformers, waste disposal units, window fans, floodlights, shaver sockets and bathroom heaters.

The appliance is connected to a fused connection unit (FCU), which is on the end of the spur cable. The spur cable should be 2.5 mm^2 twin-core and earth cable to match the cabling in the ring circuit. This means that the spur cable will still be protected by the MCB or fuse in the main consumer unit. The FCU has a small fuse in it to protect the cable to the appliance and the appliance itself. This fuse is the same type as in a 3-pin plug: for a power load up to 720 watts use a 3-amp fuse, for up to 1100 watts use a 5-amp fuse, and for up to 2800 watts use a 13-amp fuse. Always use the smallest fuse that will run the appliance as this will protect it. The cable from the FCU to the appliance should be of sufficient size to easily supply the load. If in doubt, fit 3-core 1.5 mm^2 flex.

10 Expert Points

THESE ARE SOME POINTS WORTH CONSIDERING BEFORE YOU START CONNECTING A FUSED SPUR:

1 SITING AN FCU (FUSED CONNECTION UNIT)
Mount the FCU somewhere sensible. It is the size of a single socket and should be accessible for changing/checking the fuse.

2 OUT OF HARM'S WAY
If you have children, don't mount the FCU where they can reach the switch as you can be sure they'll play with it.

3 FCUs WITH SWITCHES
It only costs a little more to buy an FCU that has the addition of a switch and a small red neon indicator. This is handy so that you can see whether the power is on and the appliance can be turned off easily.

4 FCUs WITHOUT SWITCHES
An FCU with no switch can be useful for connecting appliances, such as freezers, wall clocks and computer servers that you want to stay permanently on. This makes it impossible for anyone to inadvertently turn off the appliance.

5 BLOWN FUSES
If the fuse in the FCU blows at any time, there will be a good reason such as the wiring or an appliance faulting. A larger fuse will not solve the problem.

6 SAFEGUARDING YOUR COMPUTER
You can connect your computer into an FCU to prevent it from being accidentally turned off when you are trying to unplug something else from the multi-way extension socket behind your desk.

7 CONNECTING THE SPUR
The fused spur can be taken from the nearest socket on the ring circuit. Turn OFF the main power supply before opening up the socket. There should be two cables coming into the backing box, so there will be two reds/browns, two blacks/blues and two earths. The fused spur must be wired red/brown to red/brown, black/blue to black/blue and earth to earth. There will then be three red/brown wires in the L terminal, three blacks/blues in the N terminal and three green+yellow wires in the earth terminal when your connection is complete.

8 ORDER OF WORK
Wire up the FCU first and the connection to the ring circuit last. This way the power will be off for the shortest time.

9 EMERGENCY PROVISIONS
Keep a few fuses for the FCU with the rest of your emergency toolkit.

10 BORROWING FUSES
FCUs use fuses that are the same as those in 3-pin plugs, so you can always take one out of an unused appliance in an emergency situation.

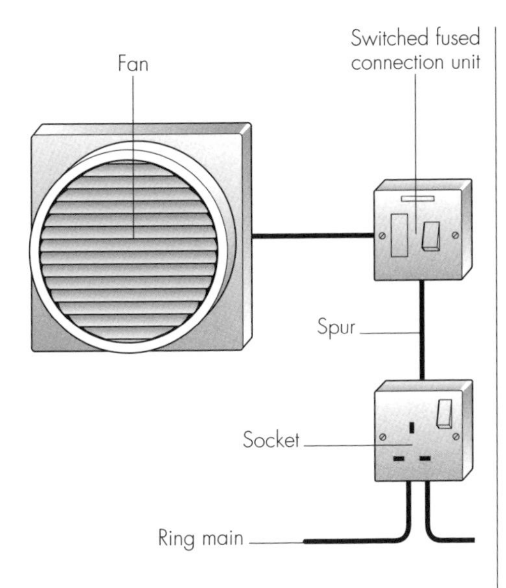

A fused spur from a socket.

A fused spur is effectively a short extension on a power circuit. The fuse protects the appliance wiring, which will probably be a smaller size than the ring circuit cable. The wiring from the appliance to the FCU will most probably be 1.5 mm^2 twin-core and earth cable. Under BS 7671 bathrooms are zoned and only SELV should be used and only suitable heaters/equipment are permitted in certain zones. The largest fuse that you can use in the FCU is a 13-amp fuse, of the same type as you'd find in a 3-pin plug, and you would have to use 1.5 mm^2 cable with this size of fuse. However, a fused spur could provide power for a small item such as a cooker hood or a bathroom air extractor fan. In this case a 0.75 mm^2 or 0.5 mm^2 3-core flex should be used to connect the appliance and a smaller 3-amp fuse must be fitted to match. The 3-core flex will have a brown, blue and green+yellow core.

The 2.5 mm^2 cable from the socket on the ring main goes up to feed the FCU. The three wires (red/brown, black/blue and earth) are connected to the "supply" terminals in the FCU.

The cable or flex from the appliance is connected to the terminals in the FCU marked "load". The live wire, which is a red/brown cable and a brown flex, must be stripped, folded double and put under the "L load" terminal (refer to "Cable and flexible cord" on page 38). The neutral wire, which is black/blue in cable and blue in flex, stripped and doubled, goes under the "N load" terminal. The earth wire goes to an earth terminal; if it is the only wire under the terminal it should also be folded double.

FCUs come in several different types; the simplest has just a fuse in a holder, while others have a switch and/or a red neon indicator to show when the power is on.

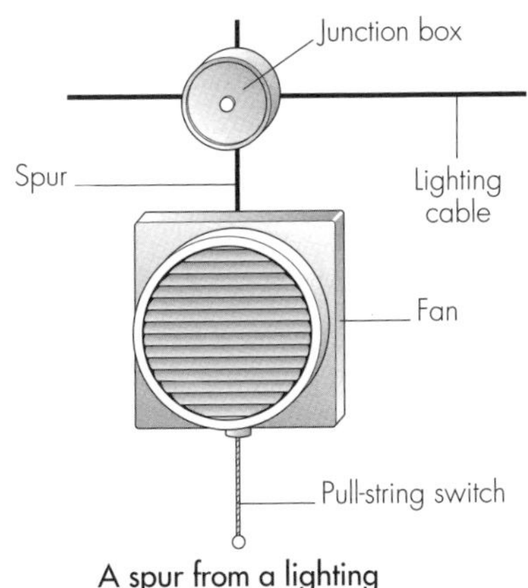

A spur from a lighting circuit via a junction box.

10 Expert Points

TEN STEPS TO WIRING A CEILING ROSE:

1 CONNECTION BLOCKS
Don't be disheartened by the number of connections in a ceiling rose on a loop-in system. Look closely and you will see there are four little brass connection blocks, which are labelled in the moulded plastic. Follow the diagram to wire them up (see page 46).

2 CONNECTING THE CABLES
The wires in the cables are solid copper core. All the reds/browns should go under the three screws in the "loop-in L" terminal block. There will be one black/blue wire with a bit of red/brown tape or a red/brown sleeving on it; this is the live return from the light switch and goes under the "L load" terminal. The blacks/blues should go under the three "loop-in N" screws. The green+yellow earth wires all go to the earth terminal.

3 CONNECTING THE FLEX
The pendant flex has a thin brown live and blue neutral wire. The brown wire goes to the "L load" and the blue wire goes to the "loop-in N" terminal.

4 EARTHING METAL
If there are any metal parts on the exterior of the light fitting, then there should also be a green+yellow earth wire going from the loop-in ceiling rose to the light fitting. The purpose of this earth wire is to protect you from electric shock in the event of a fault.

5 LABELLING LIVE WIRE
If none of the wires in the ceiling rose has a red/brown tape label or sleeving, you must put some on the black/blue wire from the switch that connects to "LOAD" in the connection block. This black/blue wire will be attached to the live wire that goes to the light fitting. The red/brown tape or sleeving is an indication that this wire will become live when the switch is turned on.

6 POWER CABLES
Remember that there is a power supply cable coming in and going out of each ceiling rose in a loop-in system; the exception to this is the last rose in the chain, which will only have a power supply coming in.

7 SUPPORTING THE LIGHT FITTING
Hook the two wires from the pendant cord over the two small posts moulded into the loop-in fitting. These take the weight of the fitting without pulling the wires out.

8 FINER DETAILS
Strip the outer insulation from the pendant cord so that it is the right length to go over the strain relief posts, but not so long that the blue and brown wires show when the rose cover is back in place.

9 USING AN ELECTRICIAN'S SCREWDRIVER
You will need a good quality, 4 mm electrician's screwdriver to fit the loop-in rose as the screws are very small!

10 USEFUL DIAGRAMS
If you are replacing a ceiling rose, make a diagram of where all the wires go before you start. Mark the wires with bits of masking tape so you can put them all back where they came from.

Lighting: loop-in circuit

The lighting loop-in system has been around for about 30 years. It is a very neat way of combining the benefits of a junction box with a ceiling rose mount. Loop-in ceiling roses are designed to make the electrical work faster as all the connections are in the one unit. However not everyone likes to see the plain, round ceiling roses screwed to the ceiling.

Before the loop-in system was used, the connection between power supply, switches and light fittings was always made in a 4-terminal junction box in the loft or ceiling space. This junction box system is still used for wall-mounted lights and other lights where there is no space in the ceiling rose for the loop-in connection block. Don't be concerned if you have a junction box lighting system because they are perfectly reliable and safe if they are properly wired up.

If you turn OFF the main power supply and look inside a ceiling rose, you will probably see about ten wires, which is typical of a loop-in system. If you find only two live, two neutral and two earth wires on just three terminals, the light has a 4-terminal junction box located elsewhere.

The loop-in fitting is a little complex at first glance, but study the diagram and you'll soon see it's quite easy to follow. First, trace the black/blue neutral wire from the power supply cable. It goes straight to the "N" connection block and feeds the outgoing neutral power cable to the next loop-in fitting, which is also connected. The neutral wire from the cord to the bulb is joined here as well and completes the neutral wiring. The red/brown live wire comes in on the same power supply cable and goes to the "L supply" connection block, linking it to the outgoing live wire and the next loop-in fitting in a similar manner.

The cable from the switch should have one red/brown wire. It also has a black/blue wire, which has a red/brown tape or sleeving on it to show that it is live when the switch is on. The red/brown wire from the switch carries the power from the "L loop-in" terminal to the switch. From the switch the power goes back to the loop-in fitting via the black/blue, red/brown-tagged wire, which is then connected to the live wire from the bulb. All the earths are gathered in the earth block.

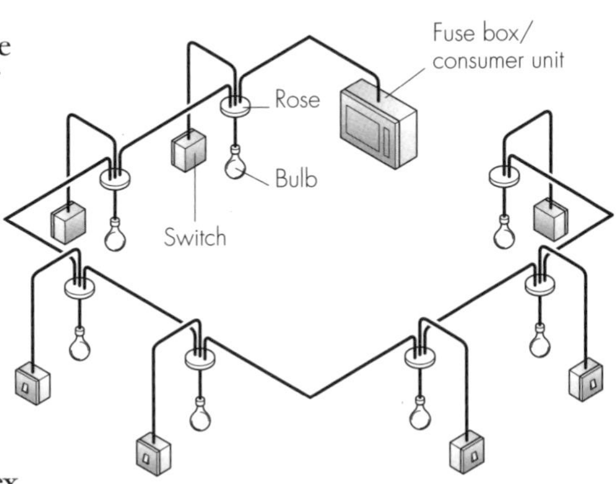

Lighting loop-in circuit.

If the light fitting has any metal parts, it must be properly earthed and there will be a green+yellow wire bolted to it. This wire must go back to the loop-in ceiling rose and be connected with all the other earth wires. If the light switches have metal faceplates, then these will also need to be earthed. A short linking green+yellow earth wire will attach the faceplates to the backing box, which is earthed via the connector in

10 Expert Points

TEN STEPS TO WIRING A JUNCTION BOX:

1 PREPARING THE CABLE ENDS
Strip back about 80 mm of the outer grey or white insulation on the cables that will go into the junction box, to give you plenty of wire to play with inside the box. Any excess wire can be neatly folded away before the lid goes on.

2 PREPARING THE WIRE ENDS
Strip back about 15 mm of the insulation at the end of each copper core in the cables. This will ensure that each core passes right under the terminal screw and is held firmly in place.

3 LABELLING LIVE WIRE
First strip 15 mm of insulation off the cores in the cable coming from the switch. The red/brown live wire carries the power to the switch and the black/blue wire carries power back to the junction box when the switch is on. You must put a piece of red/brown tape or red/brown sleeving on the black/blue core.

4 EASY CONNECTIONS
It is sometimes easier and less fiddly to take the screw right out of a terminal and put all the cores into the hole underneath it, before replacing the screw. The cores usually catch if you try and thread them under the screw.

5 SECURE CONNECTIONS
If you have put four cores under a terminal screw, then make sure you can see the tips of all four on the other side of the terminal, so that you know that they are all tightly captured under the screw.

6 LABELLING THE CABLES
Using an ordinary pen, write what each cable is for on the grey or white outer insulation. This way you can be sure which cable you are dealing with, and it will prevent you muddling them up. It's also very handy if you ever have to go back and do any other work at a later date. Use very general labels, such as "Feed in", "Feed out", "To switch" and "To light fitting".

7 ALLOWING FOR FUTURE CHANGES
If possible, leave 200–300 mm of slack on each cable that goes to the junction box; if you need to make any changes at a later date, this will make the job much easier.

8 FITTING THE LID
Make sure you fit the lid of the junction box properly so that no "gaps" are showing in the empty box notches.

9 LABELLING THE BOX
When you have screwed the lid back on the junction box, label it with correction fluid or black/blue marker to describe its function. This will make life easier if you're looking for this box in five years' time!

10 INSULATED LOFTS
All cables and junction boxes in the loft should be on top of the insulation.

the loop-in ceiling rose. Even if you fit a plastic backing box and a plastic faceplate, you should still connect the earth wire to the terminal in the backing box so that the connection is available in case anybody wishes to fit a metal faceplate to the switch in the future.

Junction box connections

This traditional method of using 4-terminal junction boxes to wire up lights is especially useful for wall lights and ceiling lights that don't have a ceiling rose big enough to contain the loop-in connectors. The big advantage is that junction box connections can be placed just about anywhere and typically in the spaces between floors and ceilings or in the loft. If you have plasterboard walls, junction boxes can also be hidden in the space between the wall battens.

Junction boxes are usually about 80 mm in diameter with a screw-on lid. Inside there are four identical brass screw terminals. A junction box

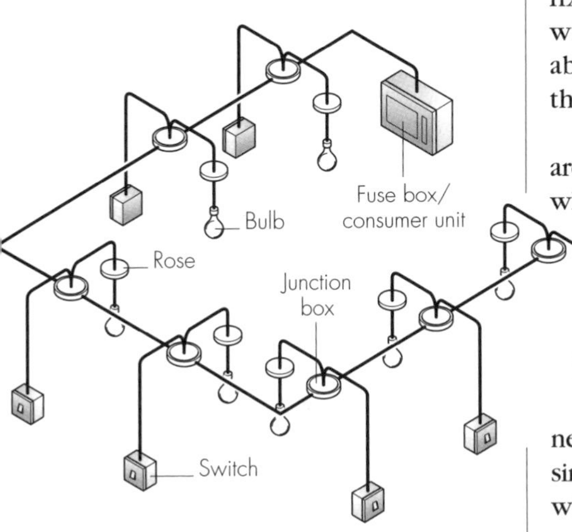

Junction box lighting connection.

has four cables going into it if it is in the middle of a lighting circuit. There will be a power cable coming in from the circuit, a power cable going out to the next light, a cable to the switch and a cable to the light itself. All of these cables should be 1.0 or 1.5 mm^2 twin-core and earth cable. All the bare earth cores must have green+yellow sleeving as normal. A junction box is just the same as a loop-in ceiling rose, the only difference being that there are four big screw terminals instead of four smaller connection blocks. Junction boxes are usually round and made of brown or white plastic. There are a couple of screw holes in the base of a junction box, which are for the purpose of securing it to a rafter or batten. Use two No.6 x 25 mm screws to attach it. If the box is to be fixed, make sure that you do it before wiring it up because you won't be able to get to the screw holes once the wiring is in place.

All four terminals in the junction box are the same, so you can use them in whichever sequence you like, although it is best to match them to your diagram. Before making any connections, you could put a splodge of red/brown, black/blue and green paint on the three terminals for live, neutral and earth, which will make it simpler to wire up. Do not use the box while the paint is wet!

Refer to the diagram on page 80 to check how the connections are made. All the live red/brown cores go under one terminal EXCEPT the black/blue live with the red/brown tag or sleeving, which goes to the terminal with the brown or red/brown live from the light fitting. All the neutrals go to one terminal. All the green+yellow earth cores go to the earth terminal.

On the lid of the junction box, you will see numbers from 2-4. These show you how to fit the lid depending on how many cables enter the box.

CHAPTER 4

Hardware

The term "hardware" is given to all the basic materials such as the cables, boxes, sockets, switches, etc., which make up the electrical system that you have in your house. As with just about everything else in life, you get what you pay for and there are very few bargains with electrical hardware. After all, if you are going to all the trouble and expense of bringing in an electrician or you are going to give up your hard-earned time to do the work yourself, it follows that you will want to have the very best hardware. Don't consider cutting costs by using low-grade fittings, and on no account use old or outdated stock, or any other cost-cutting idea that might be suggested. Just grit your teeth and buy the best that's available. Of course, this doesn't mean that you need gold-plated sockets, neo-Georgian light switches or any of the other frills and fancies that are on offer, it just means that you should get good straightforward items that conform to all the British Safety Standards.

However, there is one way you could cut costs and that's by buying all the hardware in bulk. It's very expensive to keep popping down to the local DIY store to get one of this and a couple of that. The best way forward if you are going down the DIY road is to draw up a list of your needs and decide how many metres of cable, how many light fittings and sockets, etc. - all to the same British Standard and all in matching colours and designs - that you need. Then sit down with paper and a pencil and phone around for the best prices. Ask the various suppliers to give you their very best price on the whole package and then on the different types of item, such as all the sockets, within the whole package. Working in this way, you might get the best deal buying all the sockets from one supplier, all the cable from another and so on.

Of course it's not so easy for a beginner to know what constitutes a good buy when it comes to all the fittings and fixtures because there is a fair amount to consider. There is no way of knowing at a glance how well tiny components are put together in a whole unit and whether they are made to a good standard. However, you could do your own mini survey. Once more, if you sit down with paper and a pencil and phone around - electricians, trade magazines, and even friends and family who have had electrical work done in the last five years - you should be able to build up a pretty good picture as to the best buys. For example, if two or more electricians say that a particular manufacturer makes good hardware or they always use sockets made by a particular company, then this gives you a pretty good guide. The best recommendation of all probably comes from a friend or relation - especially if they have children - who has had a system fitted in the last five years that they are still happy with. If the sockets and switches are still sound after five years' use, then those are the best products to go for.

10 Expert Points

TEN IMPORTANT THINGS TO REMEMBER ABOUT FLEX AND CABLE:

1 DIFFERENT FLEX FOR DIFFERENT PURPOSES

To a great extent, the structure and outward appearance of a flex, including whether it is insulated with PVC or rubber and covered in braided thread, is dependent on its intended use. For example, a very pliable flex that is sheathed in heat-resistant braid is generally designed for use with appliances such as irons and room heaters, which are constantly being moved from one location to another.

2 IT'S ALL IN THE CROSS SECTION

If you are at all unsure as to whether something is a flex or a cable, perhaps because they are both quite flexible and covered in white PVC, then you can sort out the confusion simply by looking at a cross section. If there are brown and blue sheathings covering the metal cores, then it's a flex, and anything that contains metal cores covered by red/brown or black/blue sheathing is a cable.

3 TWIN-CORE FLEXES

Some flex only has two conductors, a blue and brown core with no earth. They are designed for use with double-insulated appliances such as plastic hair dryers.

4 TWO-STRAND FLEX

Two-strand flex, which looks a bit like a twisted rope, is not permitted now and needs to be changed.

5 CURRENT CAPACITY

The bigger the flex, with correspondingly thicker cores and more strands, the greater the current it is designed to carry.

6 FLEX OR CABLE – THEY'RE BOTH DANGEROUS

Don't be fooled into thinking that a flex is not as dangerous as a cable. A flex that runs from a nightlight to a socket is just as dangerous as a daunting-looking cable that might run around the garage. They both carry electricity and need to be treated with equal caution.

7 PROPER CONNECTIONS

Flex and cable must never be joined by twisting cores together and binding the whole join with insulating tape; this kind of repair is not safe or reliable and is extremely dangerous. Always use the recommended connectors for flexes and junction boxes for cable.

8 AVOID OVERHEATING FLEX

As a general rule, it is not a good idea to extend indoor flexes because the cores can get overheated. If the flex from your appliance won't reach the socket, then either move the appliance closer to the socket or extend the power circuit and fit another socket.

9 SAFETY OUTDOORS

Flex for outdoor use, such as the one for the mower, is generally bright orange in colour, for the simple reason that it can easily be seen. Because outdoor flex is going to be used in wet conditions, from the rain, dew or hosepipes, it is always a necessity to fit a portable residual current device (RCD) – a cut-out device – between the plug and the socket.

10 ASK FOR EXPERT ADVICE

If you have any doubts as to the suitability of a cable or a flex, then visit a specialist supplier and explain your needs. Ask them to label the rolls of cable or flex, so that you are left in no doubt.

Cable and flexible cord

Most beginners get a bit mixed up when it comes to cable and flexible cord (flex). They tend to group them all under the general term of "wires" and/or "cables" and this leads to confusion. Start by remembering that flex refers to all the flexible cords that run from appliances to the plugs that push into sockets. So, for example, irons, toasters, kettles, TVs, etc. are all linked to the power supply by way of a flex. And, of course, flex is designed to be flexible for the plain simple reason that appliances are moved about. Once you are clear in your mind as to what constitutes a flex, then you can take it that if it isn't a flex, then it's going to be a cable. If it looks flattened in section and runs under the floor, through the loft, is fixed to a wall with clips or runs in trunking, then it's a cable. So, a flex is flexible with a plug on one end and an appliance on the other, and a cable is fixed and inflexible.

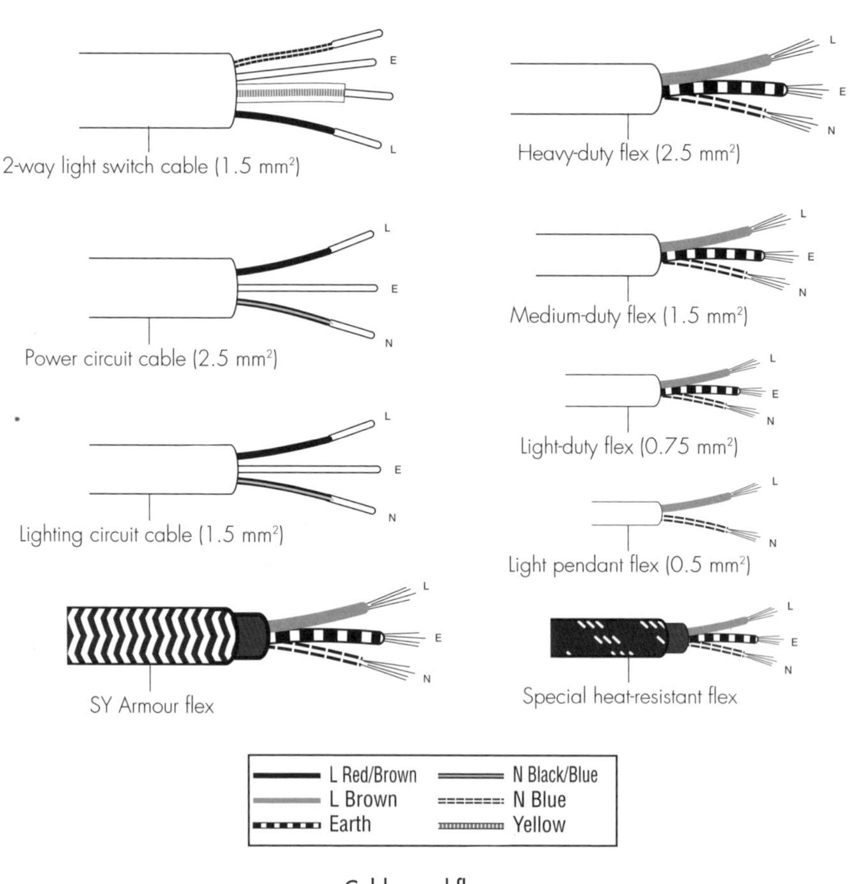

Cables and flexes.

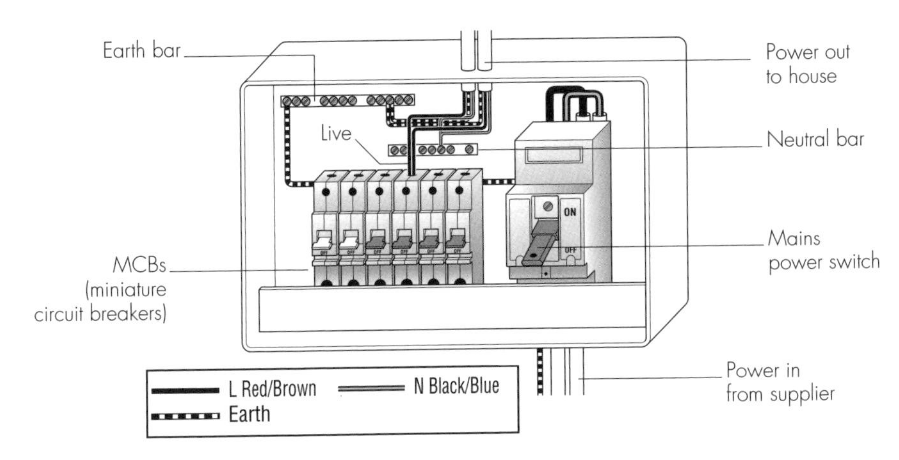

A consumer unit.

Fuse boxes and consumer units

If you look inside your meter cupboard or on the meter board, which might be in one of various places such as in the garage or under the stairs, you will see that it contains a service cable, one or more sealed fuses, the actual meter and a fuse box (consumer unit).

The electrical power comes in through the thick black service cable and then on through a black/blue neutral cable and a red/brown live cable, through a sealed fuse, through the meter and then finally into the fuse box (consumer unit).

Fuse boxes, which are usually rather large and daunting black, brown or cream boxes, are still to be found in older houses. Each of the circuits in them is protected by a massive fuse. When there is a problem on a circuit, the fuse protecting it burns out and the power is cut. When this happens, the homeowner has to identify and pull out the offending fuse block, replace the fuse wire, solve the problem that caused the fuse to blow in the first place, and then switch the power back on.

The modern consumer unit is a much more friendly device. It fulfills much the same function as the fuse box - it still protects the circuits and the power still cuts, but the fuses have been replaced with miniature circuit breakers (MCBs). Once a problem has been solved, switch the offending MCB back on.

Each MCB in a consumer unit or fuse in a fuse box is rated to protect a specific lighting or power circuit. Generally, lighting circuits are protected with a 5-amp MCB or fuse, power circuits with a 20- or 30-amp MCB or fuse and cookers with a 45-amp MCB or fuse. Fuses are usually colour-coded in white for 5 amps, blue for 15 amps, yellow for 20 amps, red for 30 amps and green for 45 amps. Some MCBs are colour-coded in the same way, but others are simply labelled with the rating.

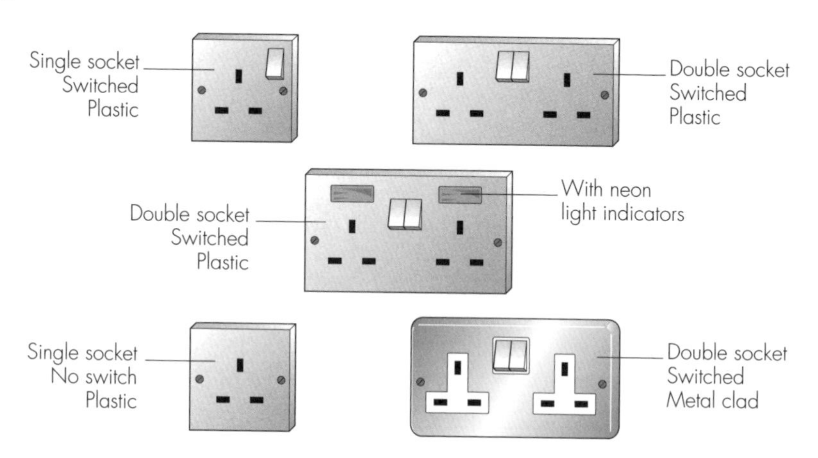

Types of socket.

Sockets

Sockets come in a vast array of designs and materials, from stainless steel to plastic and brass. You can use any socket type as long as it has British Standard approval, which is shown on the socket with a small "kite mark" symbol. As with all electrical components, it is advisable to buy the best quality product you can. Cheap sockets will work when you first buy them but tend to grip the plug pins rather weakly as time goes by, so you may end up with the socket sparking and fizzing. They are also usually a little thinner in construction and can crack when they're screwed together. A good recommendation is to buy MK products because they are reliable, last a long time and have terminals that are easy to connect.

Inside the back of a socket you will see three terminals, which should be labelled "L" for live, "N" for neutral and "E" for earth, each with a screw for holding in the wires. The screws for fitting the socket to the backing box are usually in two little clips on the back of the socket front. Remove them from their holders before wiring.

If a socket is at a low level where it might get knocked by a trolley or wheelchair, for example, you might consider metal-clad sockets and backing boxes. These are made of pressed steel and are significantly stronger than the plastic variety.

Check with your supplier where particular designs of socket can be used. There are many designs of brass- or gilt-fronted sockets, which are intended to look Victorian or Georgian in design. These can only be used in the house, where it is nice and dry, the same as ordinary sockets.

For use outside you must use only approved sockets with an IP rating, such as IP64, suitable for outdoors. These sockets are usually made of a soft, flexible plastic that won't shatter when hit, unlike the brittle sockets in the house. They often have a small door on the front that covers the whole plug and has a rubber seal all around to stop rain and spray getting in. If you want power for garden tools, the system must be protected by an RCD, which cuts off the power if there is a fault. As RCD sockets for outside use are very expensive, it would be better to put an RCD safety breaker in the consumer unit or fuse box and use an ordinary IP64 socket outside.

Fused connection units

Fused connection units (FCU) are designed to supply power to appliances that are permanently connected to the electrical system. They are the same size as a single socket and can be fitted to a single socket backing box, either sunk in the wall or mounted on the surface. Ideally they need a fairly deep backing box, say 40 mm, because there can be quite a few wires in the unit, and the faceplate, with the fuse, is quite thick and takes up a lot of room in the box.

The cable or flex to the appliance can be connected to the FCU in several ways (see page 95). However, some FCUs have a clamp inside to prevent the flex being pulled out. The clamp is usually a small nylon strip with two screws holding it down. The flex to the appliance comes out of the FCU through the front or edge of the faceplate, depending on the model.

There are several different types of FCU, the most common of which is a simple unit with a small fuse holder visible on the front. The fuse holder is usually held in with a small screw. The fuse, which will be 13, 5 or 3 amps, protects the cable or flex to the appliance, since this will usually carry much less current than the 1.5 mm^2 cable used in the power circuit. Without the fuse, the cable or flex to the appliance could overheat and melt if the appliance malfunctioned.

Other types of FCU include an on/off switch and a small red indicator light. The switch is very handy for devices, such as smaller immersion heaters or kitchen extractor fans, which you may want turned off for a while. It's not usually a good idea to use a switched FCU for a doorbell transformer, chest freezer or computer server, because the power could easily be accidentally turned off. The little power indicator light is handy because it shows if power is going to the appliance.

They are available in thermoplastic, which is a brittle material, and that is fine if the unit is going to be positioned where it can't get knocked. If the FCU is going to be used in a more industrial environment or anywhere where it may get damaged, then always use a metal-clad FCU and a similar backing box. Metal-clad units are made from pressed steel and can take quite a bit of abuse. However, remember that FCUs are only suitable for indoor use unless they have an IP55 or better rating.

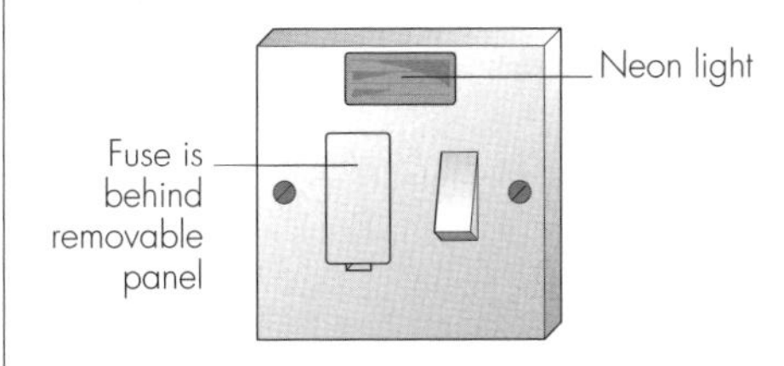

A fused connection unit with switch and neon light indicator.

Double-pole switches

When an appliance requires a current of more than 3 kW, and therefore a 13-amp fuse, an FCU cannot be used because the largest fuse that these units will take is a 13-amp. You will therefore require a double-pole switch for appliances such as storage heaters and larger immersion heaters to allow you to switch the circuit off. They can be mounted on a normal single surface-mounted backing box or in a single metal, flush-mounted box, both with a minimum depth of 35 mm. They are available with or without a red neon indicator.

Small double-pole switches carry 20 amps, which will safely switch appliances up to about 4.5 kW; you will see one of these next to a storage heater, for example. The incoming power cable is usually 2.5 mm² and comes from a 20-amp MCB or fuse in the consumer unit or fuse box. The cable going out to the appliance can pass either through a small hole in the faceplate or through the backing box. The flex to the appliance should be 2.5 mm² and must be clamped in the double-pole faceplate using the two screws and the plastic clamping strip.

Double-pole switches are also available in 32-, 45- and 50-amp ratings. A 32-amp unit will still fit in a single 35 mm-deep backing box, but 45- and 50-amp units need a double socket backing box. It is possible to order a 45-amp double-pole switch to fit a single socket box but they cannot be bought off the shelf.

Wall heaters or electric showers in bathrooms and shower rooms must have a double-pole switch that is mounted on the ceiling and activated with a pull-cord. Ceiling-mounted switches are available in 16- and 45-amp ratings. These switches must also have a flag indicator, which is a small plastic tag that pops up to show whether the switch is on or off. Although a ceiling switch may also have a red neon indicator to show whether it is on or off, the flag indicator must be there in case the neon light fails. Ceiling switches are normally surface-mounted on a single 35 mm box, but they can be fitted on flush boxes as well. Ideally all double-pole switches should be clearly labelled to describe the appliance they are connected to. Small plastic engraved labels, which are self-adhesive and have a peel-off backing, are especially suitable for the purpose and are available from electrical wholesalers.

Cooker switches

Cooker switches are just like normal 32- or 45-amp double-pole switches, which means that when the switch is turned off both the neutral and live wires are cut off and only the earth wire is still connected to the cooker. You need a 32-amp switch for a cooker requiring less than 7.5 kW of power and a 42-amp switch for one requiring more than 7.5 kW. You can find out how much power your cooker needs by looking at the rating plate on the back of the cooker. Unlike double-pole switches, cooker switches are pre-labelled and some models also have a 13-amp socket.

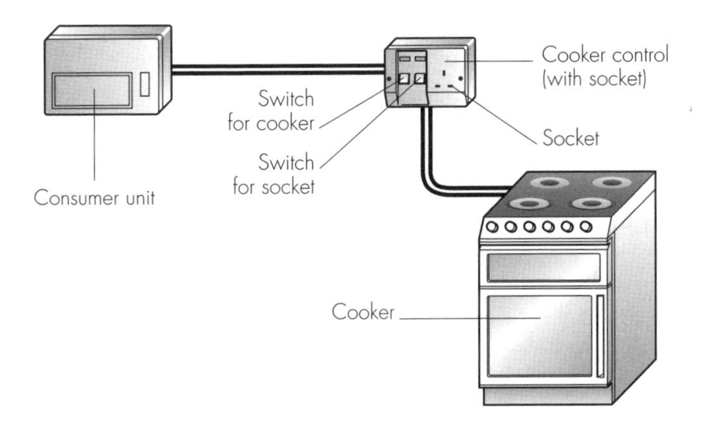

A cooker switch (each item is shown at a different scale for clarity).

Older style cooker switches are larger than a normal double-pole switch and use a backing box that measures 160 x 115 mm and is 45 mm deep. The deeper box allows room to accommodate the heavy 6 mm^2 cable needed for a cooker, which does not bend easily. Cooker switches have two sets of connections on the back of the faceplate - live, neutral and earth supply, and live, neutral and earth load. Both the supply and the load cable must be 6 mm^2.

The load cable may go to a cooker connection unit (CCU), which is essentially a square 45-amp junction box with three terminals inside. The CCU fits onto a normal single backing box, either flush-fitted or surface-mounted. The backing box should be at least 35 mm deep.

More modern cooker switches are designed to fit onto a normal double socket backing box, although this would need to be 35 mm deep to accommodate the heavy 6 mm^2 cable. The cooker switch should be clearly labelled "Cooker" and should be mounted to the left or right of the cooker, NEVER above it in case of a cooker fire. There are various alternatives available: a cooker-only switch, which is all most people need, or one with the addition of a 13-amp socket, which might be handy for something like the electric kettle. The switches are also available with a red power indicator light for the cooker switch and/or the 13-amp socket. The 13-amp socket is available with, or without, a switch of its own.

Remember that, as well as the cable and the cooker switch, the MCB or fuse in the consumer unit or fuse box must be the right size for the cooker. To calculate how much amperage will flow through the circuit, divide the cooker wattage on the label plate by 230 to give you the number of amps and then round the number up. For example, if the cooker requires 7.5 kW of power, the wattage is 7500; divide that by 230 and you get 32.61 amps, so you need a 40-amp switch.

Shaver sockets

Under BS 7671 bathrooms are zoned and only SELV should be used and only suitable heaters/equipment are permitted in certain zones.

Electric shaver sockets are available in two types: shaver socket power point and shaver power supply units. There is one vital distinction between these two types and that is safety.

A shaver socket power point is just a socket connected to the lighting or power circuit that has a small fuse in it and continental-style 2-round-pin connectors. They fit on to a normal single socket backing box, either surface-mounted or flush-fitted. However, this type of socket can only be used in bedrooms where there is no sink or shower as there is a high risk of electrocution if it gets damp. If a device other than a shaver is plugged into this type of socket, the small fuse will blow straight away.

Shaver power supply units are specifically designed for bathrooms and shower rooms. This type of unit is extremely safe and often has sockets on the front for both 110 volts (American) and 230 volts (European). The unit provides an isolation transformer between the shaver and the mains electrical supply, which means that it is totally disconnected from the mains power. This includes the earth wire, so even if there's an earth fault elsewhere is the house it will not extend to the shaver. On the shaver side of the isolation transformer there is a trip or cut-out that will activate if there is a fault in the shaver. The cut-out will also activate if a device other than a shaver is plugged into a shaver unit.

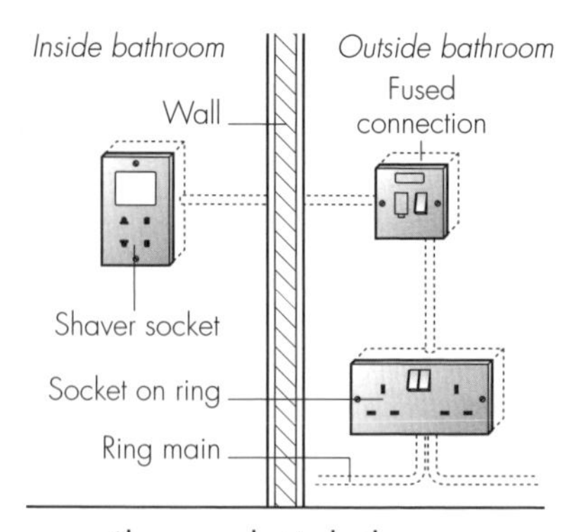

Shaver socket in bathroom.

Most shaver power supply units are designed to fit onto a double socket backing box, with the box mounted portrait-style. The backing box can be either surface-mounted or flush-fitted and must be 47mm deep because the transformer and other parts of the shaver supply take up a lot of space.

If a shaver power supply unit is going to be mounted next to a mirror, it may be worth considering using a combined shaver power supply and strip light. The shaver power supply unit would fit over a flush-fitted box and the strip light would be surface-mounted over the mirror.

Light switches

In any house you will see at least three types of light switch – usually a pull-cord in the bathroom, single-way switches in the bedrooms and two-way switches in the halls or stairways.

A two-way switch is one that operates a light from one position in conjunction with another two-way switch that works the same light from another position, for example, at the top and bottom of a flight of stairs. The advantage of this type of switch is that you won't be left groping for the switch in the dark when you enter a room or corridor from the end furthest from a switch.

A two-way switch is easy to identify by looking at the back of the faceplate, where you can see three connection terminals. The terminals are labelled "COM, L1" and "L2", where "COM" means common, and "L" means line. A special cable links a pair of two-way switches. It has four cores, is described as three-core and earth, and is usually 1.0 mm^2 in size. Three of the cores are sheathed in red/brown, blue and yellow, but the fourth earth core is bare and must have green+yellow sleeving slipped onto it.

If you wish to add an extra switch to a two-way system, maybe because a corridor has three or four doors opening on to it, you will need to install some intermediate switches. These switches have four terminals, plus an earth. In an intermediate switch the two pairs of terminals are labelled "L1" and "L2". All of the above switches generally fit into 25-mm-deep backing boxes, which can be flush-fitted or surface-mounted as required.

Ceiling pull-cord switches must be used in bathrooms, washrooms and shower rooms to avoid the danger of electrocution. They work in the same way as wallplate units and have similar terminal layouts. They are available in both single-way and two-way configurations, so if you have an en-suite bathroom with two doors you can fit a lighting pull-cord switch at both entrances. Ceiling pull-cord switches are usually surface mounted on a shallow round backing box. However, the backing units are not universal fittings like socket boxes so you will have to purchase the box and switch at the same time to be sure that they will match. Pull-cords are normally fitted with a rather dull white knob, but there are many fancy metallic, ceramic and wood fittings available in DIY stores that can be used instead to complement your bathroom decoration.

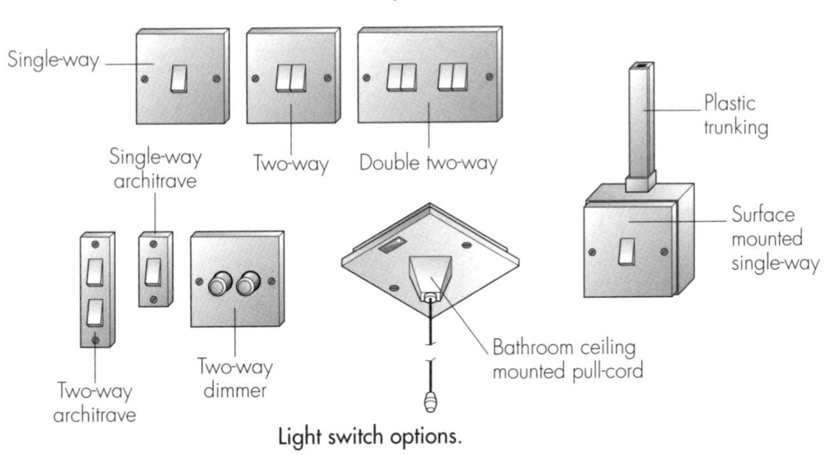

Light switch options.

Ceiling roses

Ceiling roses are divided into two types depending on the kind of lighting circuit used in a particular room. The round white ceiling roses that have a short flex and bulb holder hanging from them are probably on a loop-in system. Loop-in ceiling roses are commonly fitted in modern houses because they are very quick and easy to install, and cut down on wiring and junction boxes. To make sure whether a ceiling rose is on a loop-in system, first turn OFF the main switch and then unscrew the ceiling rose cover. Look inside at the number of small connection blocks with screws. A loop-in rose has four connection blocks, each with about three screws. If the ceiling rose has only three connection screws, it is connected to a four-way lighting junction box elsewhere, probably in the roof space or under the floorboards above.

The connection blocks in a loop-in ceiling rose are labelled as "Earth", "Loop L", "Loop N" and "COM". All the green+yellow earth wires go to the earth terminal block. The red/brown live and black/blue neutral wires in the cable from the consumer unit or the previous light fitting on the loop go into the loop terminals. The red/brown live and black/blue neutral wires in the cable going to the next light in the loop are also connected to the loop terminals. The light fitting flex is looped over the two small "ears" so that the weight of the light and shade is not hanging on the screw terminals. The blue neutral in the flex from the light fitting goes to the loop N terminal and the brown live goes to the COM terminal. The red/brown live wire in the switch cable is connected to the loop L and the black/blue neutral wire with red/brown sleeving goes to the COM terminal.

The ceiling rose base has a couple of screw holes for mounting it to the ceiling. The fixings that are needed to hold the rose up depend greatly on the weight and type of the light fitting, as well as on the material of the ceiling itself. Most ceilings are plasterboard, which will not support a lamp weighing more than a few kilos.

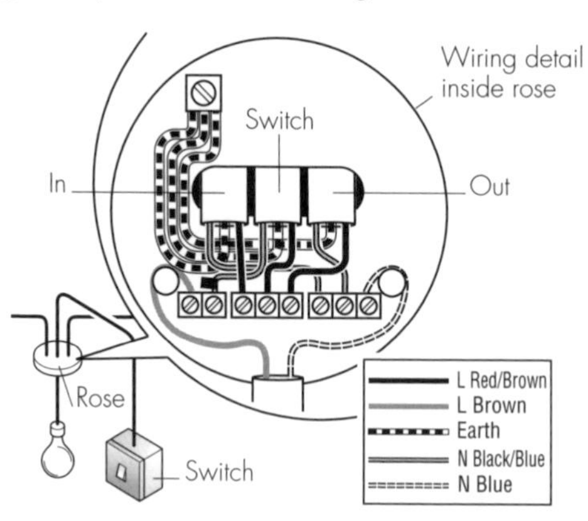

A loop-in ceiling rose.

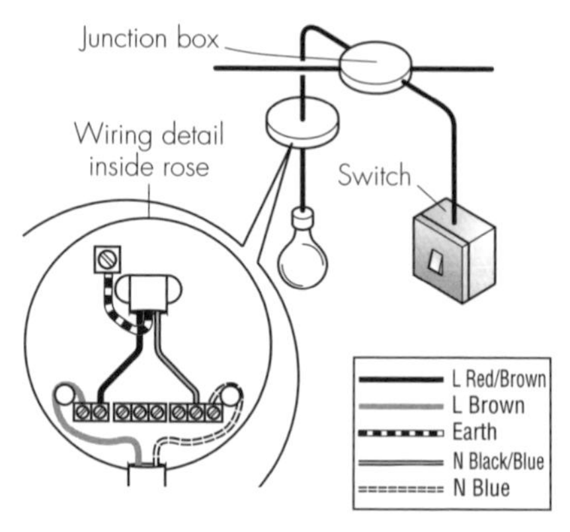

A Junction box lighting system.

In this case, the ceiling fitting should be positioned underneath a joist in the ceiling space. A couple of No.8 x 80 mm cross-head screws driven into the joist will support a lamp weighing up to 5 kg. For heavier light fittings, threaded bar and plates bolted through the joist will have to be used.

Lamp holders

There are dozens of different lamp holders, ranging from the ornate chandelier type to the plain white pendant fitting, as well as fluorescent strip lights and those suitable for low voltage outside lights, underwater lights and floodlights. Most lights are designed for indoor use, so special attention needs to be paid to any unit that is to be fitted outdoors. It will have to withstand all weathers and will need to be specially sealed to avoid water tripping out the lighting circuit.

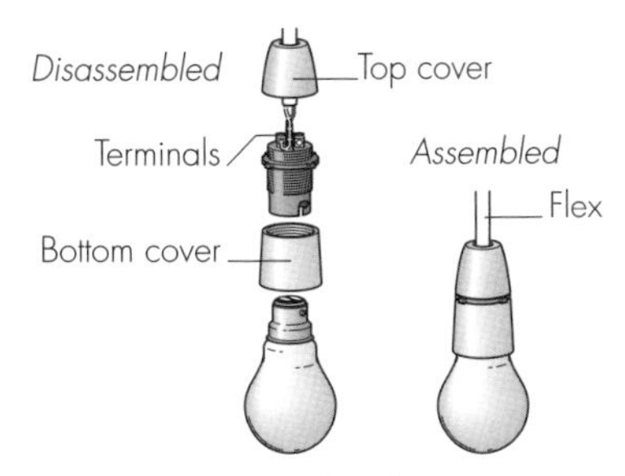

A pendant fitting.

The simple white plastic pendant fitting is the type seen hanging from many a ceiling with a lampshade or light fitting attached. The shade is held onto the pendant with a small ring or collar, which comes already screwed onto the fitting. The pendant is supported by a short flex tethering it to the ceiling rose, so the rose must be fixed securely. If the light fitting is made of metal, it must be earthed to prevent accidental electrocution and so it may well be necessary to change the flex supplying the fitting to one with an extra green+yellow earth core.

Most of the more fancy lamp holders that can be bought in DIY stores are pre-wired to a top section that needs attaching directly to a three-way connection block in the ceiling or wall. The wires will be blue (N), brown (L) and green+yellow (E).

Fluorescent strip lights have a similar connection block, again with an earth wire that must be connected. They have a small box in the back called "the ballast", which is a special type of transformer and circuit. Strip lights are usually pre-wired and can look quite complex inside. However, the three wires to connect will all be labelled, "L", "N" and "E". Some strip lights have a small round plastic cartridge called a starter fitted through a hole in the side of the casing. The starter contains a unit that turns the fluorescent tube on initially, after which the ballast takes over. Starters do not last forever, but the push-and-turn fitting makes them easy to take out and replace.

Low voltage lights, which have small 12-volt bulbs, are often used in kitchens and gardens because they are much safer than 230-volt equipment. Low voltage systems have transformers that drop the voltage from 230 to 12 volts and they are usually connected to a 3-terminal connection block similar to a fluorescent strip light connection.

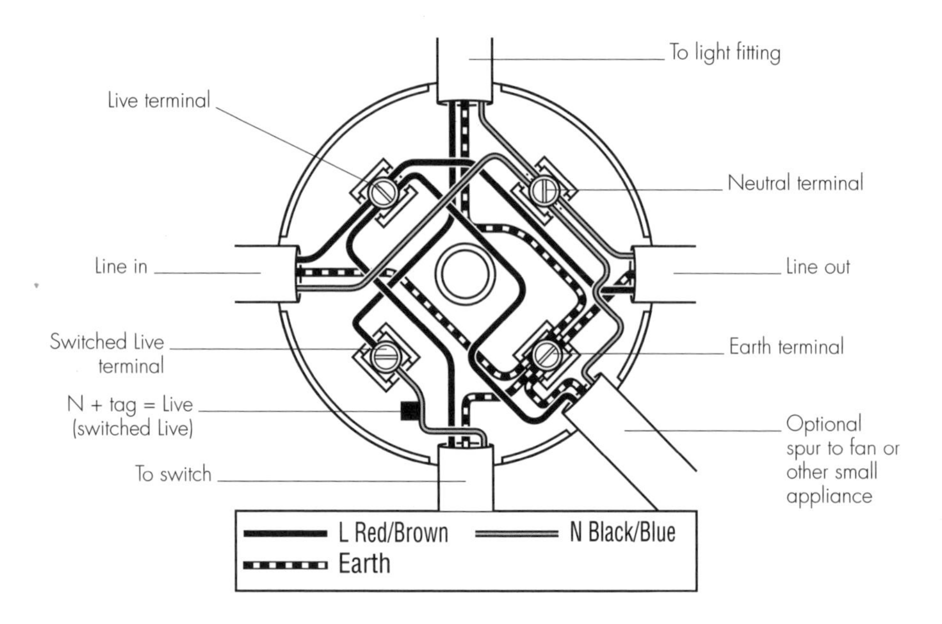

A lighting junction box with an optional spur.

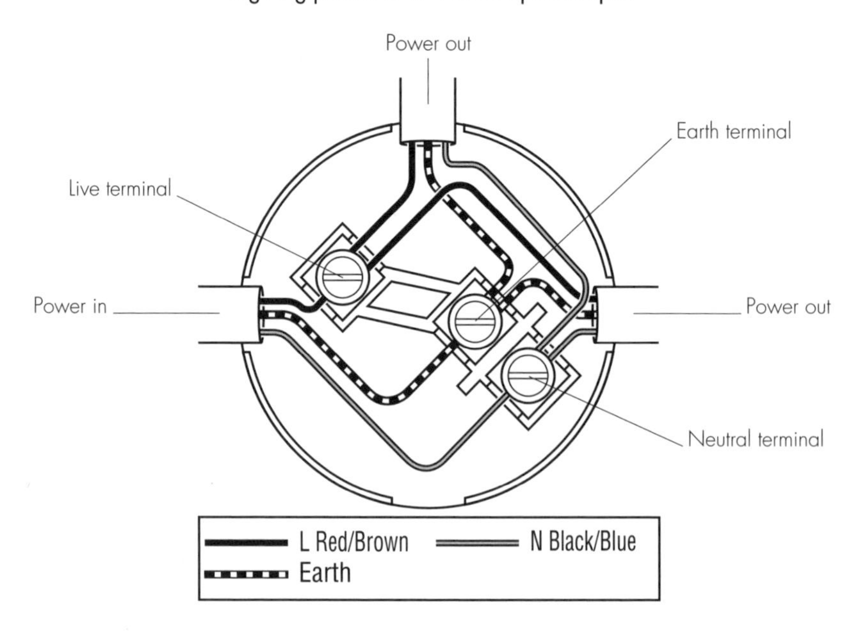

A cable jointing junction box.

Junction boxes

Junction boxes are usually round with lids that are attached with a screw, although rectangular ones are also available. The function of a junction box is to join the ends of cable together. Junction boxes come in a range of sizes to suit the size of the cable and load they will be carrying.

Junction boxes used to wire the switch, cable and light fitting on a junction box lighting system have four terminals inside. Other boxes have three terminals for connecting the live, neutral and earth wires. For lighting circuits, 5- or 10-amp junction boxes are used and for power circuits, 20-, 30- or 45-amp boxes are used. On the junction box lid you will see numbers 2, 3 and 4, which indicate how you should fit the lid depending on how many cables enter the box. For example, if you have four cables, you will need to line up the lid so that each cable is under a number 4. For three cables, you will need to fit the lid so that a number 3 is over each cable, thus sealing off the fourth, unused, hole in the box automatically.

Conduit

Conduit to carry cable is available in many materials, shapes and sizes. Oval or round section plastic conduit is usually buried in the walls and makes it easy to replace cables if necessary.

Surface-mounted conduit is either the plastic rectangular type with a clip-on lid or the round type, which has circular connecting boxes at both ends. The rectangular conduit is only for use inside the house, in the dry. It usually comes with a peel-off sticky backing for fixing it to the walls, although it is better to use a few screws as well.

The round type of conduit is available in plastic or galvanized steel and is designed for outdoor use in all weathers. The connecting boxes have screw holes in the back for fitting to the wall. Galvanized steel conduit is vandal resistant and can take a heavy blow without the cables inside being damaged, so it is ideal for using along the bottom of outside walls or in workshops. It also needs to be threaded at the ends, for which a threading machine is needed. The plastic exterior conduit is very similar to the steel type, but the ends are glued into small threaded fittings that go into the round connecting boxes, so a threading machine is not necessary.

Light bulbs

Not so many years ago we had the standard push-and-turn tungsten filament bayonet bulb in clear or pearl glass, and that was just about it. We can now choose from an enormous range of bulbs with both large and small bayonet fittings, large and small screw fittings, in just about every shape and colour you can think of. Just to list a few of the options, there are small globe-shaped bulbs, candle-shaped bulbs, bulbs with reflective sides, small halogen bulbs, white bulbs, coloured bulbs and single cap tube bulbs.

One very worthwhile revolution in the last few years has been in the development and use of energy-saving and low voltage bulbs. You can now light your entire house with low voltage systems and the actual bulbs are so small and/or attractive, that they can either become decorative features in their own right or may simply be hidden from view.

Light tubes

While most people don't really need or want to know the technical difference between a tungsten filament bulb and a fluorescent tube, they can of course easily identify a tube by its tubular shape. As far as lighting in the house is concerned, tubes come in three common sizes of 38 mm, 25 mm and 15 mm diameter, with lengths ranging from approximately 2400 mm through to small items at about 100 mm. In addition to straight tubes, there are also circular fluorescent ring tubes at 300 mm and 400 mm across the diameter of the ring. There is now also a cross-over fluorescent tube in the form of a tight S-shape. These are described as long-life bulbs and have a push-and-turn bayonet or a screw fitting, which can be used as a ceiling rose light or as a wall light, instead of a normal bulb.

Fluorescent lighting is most often used in kitchens, bathrooms and workshops where a strong direct light source is required. In bathrooms and bedrooms fluorescent tubes are popularly used with a two-pin shaver socket unit fixed above a mirror. In kitchens, and sometimes bathrooms, ring tubes fixed to ceilings are popular.

Low voltage lights

Low voltage lights are generally halogen, which gives a bright, crisp colour. Halogen colour rendering is particularly good compared to normal fluorescent and tungsten bulbs. Low voltage lights are very clever, in that each light or small group of lights has its own transformer that cuts the 230 V mains supply down to 12 V. However, the need for a transformer and additional maintenance does add to the price of low voltage lights. Because the bulbs are so small they give off a lot of heat, so it is advisable to keep any flammable objects at least a metre away from the lamp.

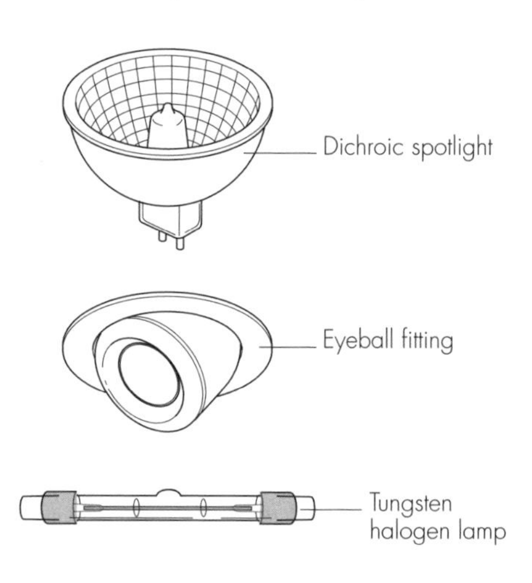

Low voltage lighting.

This revolutionary development in the size and structure of light bulbs has completely changed the way in which lights are used in the home. Once upon a time we were limited to having a bulb hanging from a central ceiling rose or a number of wall lights around a room, in shapes very heavily influenced by such antique models as Victorian oil lamps and Edwardian gas lights. Now we can choose to have strings of miniature lights that look to a science fiction future with forms ranging from flying saucer shapes through to minimalist discs. Once we might have been limited to fabric shades with fringes that created a folksy or at best Arts and Crafts look that dictated the design of a room. Now we can use lights which are so small and discreet that we can think about the effect of the light without having to dwell on the actual shape of the structure holding the bulb. Although you still have to have a lighting circuit, you can, if you choose, do away with the business of having ceiling roses and wall brackets, and go for lights that are recessed and hidden away within the structure of the building. In this way you can have pools or shafts of light coming up from the floor or down from the ceiling without the fittings themselves being on view.

TV aerial sockets

Every household with a TV has the same basic set-up. If you have a close look at the back of a basic stand-alone TV, you will see that it is connected to a large, rounded black, brown or white cable, known as a coaxial cable, and a flatter white or black electrical flex. If you now trace the cable and flex away from the TV, you will see that the cable runs to a square wall plate or socket and the flex has a plug, which is pushed into a power socket on the end.

Now unplug the TV from the mains so that it is safe to handle and have a look at the coaxial cable which has a round male plug on the TV end and a round male plug on the end that pushes into the wall socket. If you unscrew one or other of the two plugs, you will see that the cable has four structural elements. There is a thick central wire or core, an inner insulation tube around the core, an outer mesh of wire known as a screen that runs like a sleeve over the inner insulation and a sheath of outer

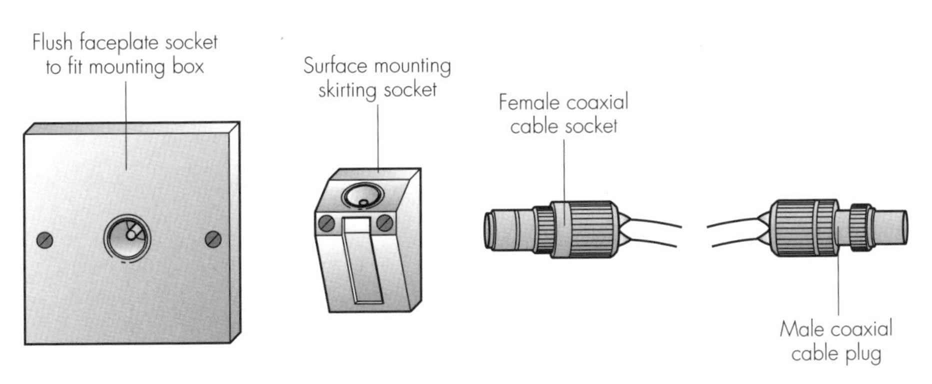

Television aerial sockets and cable plugs.

insulation. Look closely at the way the layers that make up the coaxial cable are arranged within the socket and you will see that the inner wire core and the wire mesh act as two conductors, just like two cores in an electrical cable or flex. They are kept apart by the insulation, so that they never touch.

If you were to turn off the main power supply, open up the wall socket and inspect the coaxial cable there, you would see that one end connects to the back of the socket in much the same way as it is linked to the female and male sockets. The other end runs through the roof or floor space up to the outside aerial.

While coaxial TV cables are, in many ways, quite different from electrical cables, the techniques involve in extending and replacing TV cables are much the same as those involved in extending and replacing electrical cables. So if you are skilful enough to wire up an electric socket or two, then there's no reason why you can't replace or extend your TV aerial cable.

WARNING: think carefully before you go scrabbling about on your roof. If you have doubts about your head for heights, get a professional to fix the TV aerial to the roof and to run the coaxial cable into the roof space, and then you can take it from there.

Telephone outlets

If you have a close look at the basic modern telephone system, such as a modern digital cordless telephone complete with its base station linked directly to a single master socket, you will find that there are two flexes. There is an electrical flex running from the base station to a transformer plug that is pushed into a power socket, and there is a flex running to the telephone company's master phone socket. If you have one of the old, enclosed junction boxes, rather than a socket, now is the time to ask the telephone company to change it to a new one. This is important because you can't run extensions from an old junction box.

Although not so long ago it was the norm to have a single phone in the hall and, if you were very grand, a telephone in the bedroom, nowadays most people have phone sockets all over the house, maybe for a fax machine and a computer. The other change has been to the rules and regulations that govern fitting extensions. The present law says that you can make changes on your side of the master socket. The phone company owns the pole in the road, the cable running from the pole to your house and the master socket. So you must not open the master socket but you can connect a telephone plug and wire up new extensions that are linked to the master socket.

You may want to extend your telephone systems for a single telephone in the hall to give you sockets all over the house, maybe in the bedrooms, for the computers and in the outside workshop or office. Start by visiting one of the large DIY stores and see what sort of equipment is on offer. You will find all manner of kits for everything from a single converter plug that allows you to run two phones off the master socket or a roll-out extension cable that allows you to run an extension off your master socket to another room, to kits for sockets running off a central junction box or a number of sockets running in series. In many ways the series option is the simplest to understand and to fit. If you look at a typical series kit you will see that it includes a plug to connect to the master socket, a length of cable to run from the plug to the first extension socket, another length of cable to run from the first extension socket to the next, and so on to the last socket in the line. The average kit will contain sufficient extension sockets to suit your needs, cable, little clips or cleats to fix the cable to the skirting, and an insertion tool that enables you to fit the cable to the sockets and plugs. So it is all pretty easy to manage - much the same as electrics, only somewhat smaller and more fiddly.

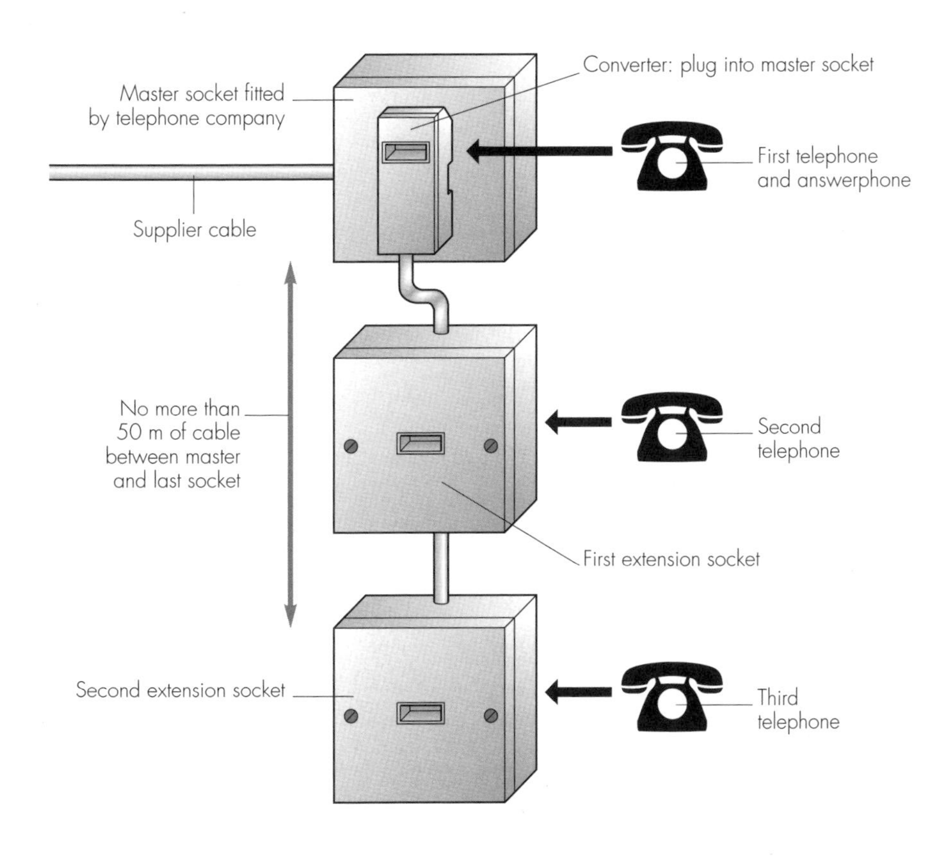

Telephone master socket and extensions (cable not shown actual lengths).

CHAPTER 5

Projects

Stripping cable

Before you start

To undertake any electrical work it will be necessary to master the technique of stripping cables and flex. Of course, there are a few tricks of the trade, which make it easier, and they are described in detail in this next section. You will need some basic tools but it is not necessary to spend loads of money on them as good quality gear is now cheaper than it's ever been in the large DIY stores.

It's a good idea to get a few odd bits of cable and flex to practise on. You don't want to be learning on the 100 mm tail end that is just poking out of the wall and must be done right the first time or you might find you have to replace the whole cable run back to the consumer unit or fuse box. Buy a metre of the different types of flex and cable from a DIY store or better still chop a few bits off some of the old stuff you might see in builder's skips (when you've asked permission) as it's ideal for practising on. As with all electrical and DIY work, take a bit of time out to practise, in good light, on a big table - when the kids are in bed is best or they'll be into everything!

Tools and materials

First you will need a few tools. Choose ones with a tried and tested brand name - after all that's what an electrician does. Go to a large DIY store, where they have bulk buying power, and the prices should be keen. You will see many brands of each tool. Have a touch and a feel before you buy - good tools feel nice and solid and will usually have a brand name you have heard of.

Pliers are the most important tool for stripping cable and flex. With practice, they will become the only stripping tool you will need and a good pair will last your lifetime - or at least until someone borrows them and never gives them back. Pliers have square metal ends to grip with and a set of cutting edges set back about a centimetre from the end; the handle grips have rubber sleeves to prevent electric shock. The grips are usually marked "1000 V CE', which means that they will stop a 1000 V shock if you cut a live wire, although you must NEVER test this out!

Side cutters are similar to pliers to look at but they have a cutting edge that is about 20 mm long and the ends do not grip. They are good for cutting cable or flex from a roll and are designed to cut using one hand with minimum effort, whereas pliers will need a two-handed squeeze on a heavy cooker cable.

Wire strippers are a bit of a luxury item. They are used for stripping the individual cores within the flex or cable, but they cannot easily strip the outer insulation.

The final item you will need is a retractable craft knife. There are lots of brands, but the recognizable brand names can be trusted; they are solid, cheap and feel good in the hand. Don't get a plastic-handled knife with the snap-off strip blades, which are not robust enough for this job.

Method

Stripping ordinary grey twin-core and earth cable is awkward until you know the trick of using the earth core to split the grey outer insulation. Never use a knife to cut off the outer insulation as you will inevitably nick the inner coloured covering or the cores. This leaves a weakness in the core wiring and, at worst, a bare bit of copper waiting to cause a short.

Push the corner of the square ends of the pliers into the end of the twin-core and earth cable and grip the tip of the bare earth core, which is usually the centre core. Then, holding the pliers in one hand and the cable in the other, pull the core down the side of the cable to slit the outer insulation, rather like undoing a zipper. Pull the core down until you are about 5 mm past the point where you want the insulation removed. Peel the other coloured cores out of the outer insulation, like peeling a banana. Pulling the cores one way and the outer insulation the other, use pliers or side cutters to cut off the outer insulation.

The insulation can be easily stripped from the individual cores of the cable with a pair of wire strippers. Some strippers can be set to the thickness of the insulation. Others have a range of notches to use for different cables, which are usually labelled with the cable size, for example, set to the 2.5 notch the strippers cut 2.5 mm^2 cable. If you look carefully, you will see that the size of the cable is printed or stamped into the outer insulation. All strippers work in the same way; they cut into the insulation round the core, but not through it, so the excess insulation can be pulled off without damaging the core. Take care that the core is not nicked because this creates a weakness or fracture point where the wire will break when bent a few times.

There are several ways of stripping flex. It is more difficult to strip than cable because there is no bare earth core to slit the outer insulation. Use the retractable craft knife to gently make a shallow cut around the outer insulation; the cut must not go right through the insulation or you will nick and damage the coloured cores. Normally only the lightest touch of a new sharp blade is needed. Next, bend the flex sharply on the cut line, you will see the cut widen until the coloured cores are visible. Turn the cable round by 90 degrees and bend it again to widen the split a bit further round the insulation. Repeat this until the end of the insulation has become detached. Grip the main length of flex in one hand and the outer insulation on the other end just below the cut point in the other hand. Then, pull the insulation off the coloured cores – a bit like pulling off a tight sock. You may need to pull the insulation off with pliers if it is stiff, but take care not to nick the cores or you'll have to start again. The inner cores of flex can be stripped with wire strippers in the same way as for cable. However, the cores in flex are made up of many fine wires so extra care is needed if they are not to be damaged.

Wiring a plug

Before you start

Wiring a three-pin plug is a skill that everyone should have even though the law now states that every new appliance must have a factory-fitted plug. Plugs can smash when dropped and flex gets damaged, so eventually you will need to change a plug. New plugs come with a small card label pushed over the pins, which shows how the plug should be wired; it's a good idea to keep one of these in your toolbox as a quick reference.

Plugs can be made from different materials, but the type made of soft-looking plastic are more robust than the ones that look like they're made of china. These softer thermoplastic plugs, made of PE, ABS (both types of polymer) or nylon, are much longer lasting and don't shatter when dropped on a hard surface.

On some plugs the lower two pins are partially insulated. These plugs are safer to handle when they are being plugged and unplugged because there is no live pin that can be touched accidentally.

Tools and materials

You will need the following tools: pliers, side cutters, cross-head screwdriver, 4 mm flat screwdriver, 6 mm flat screwdriver, retractable craft knife, wire strippers.

Method

Work on a worktop so that you don't lose any of the tiny components or stab yourself if the screwdriver slips. To remove the top from the plug, start by undoing the large screw that's in between the pins. This screw may have a cross-head or a slot, so use the appropriate screwdriver. Do not use the small 4 mm screwdriver for this large screw, which will snap the corners off the end of your screwdriver blade.

Take off the back cover. You'll see the three pins, a fuse and a small strap gripping the flex. Some plugs have small plastic jaws gripping the flex, but this type is not very common. Unscrew the two screws that secure the cable grip and remove it altogether. Then, using the small screwdriver, loosen the three small brass screws in the pins until you cannot see them showing in the holes where the wires go, so that the wires will push in more easily. Lay the screws carefully to one side.

Mark around the outer insulation of the flex around with the knife and strip off about 50 mm, leaving the three coloured cores exposed. Measure how long each core needs to be by laying them against the plug. To measure accurately, make sure that the cord grip will be holding the outer insulation, then lay each core wire to the correct pin and mark how long it needs to be in order to pass through the terminal holes. Cut the cores at the points marked with the side cutters. You will find that each core is now a slightly different length, with the green+yellow earth the longest and the brown live the shortest. Using wire strippers, strip off 6 mm of the insulation from the end of each core.

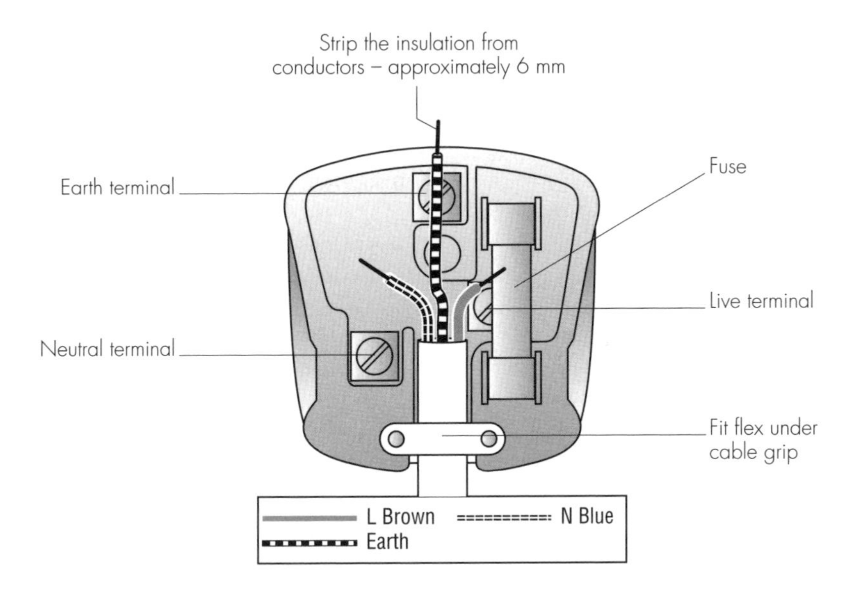

Wiring a plug: Step 1.

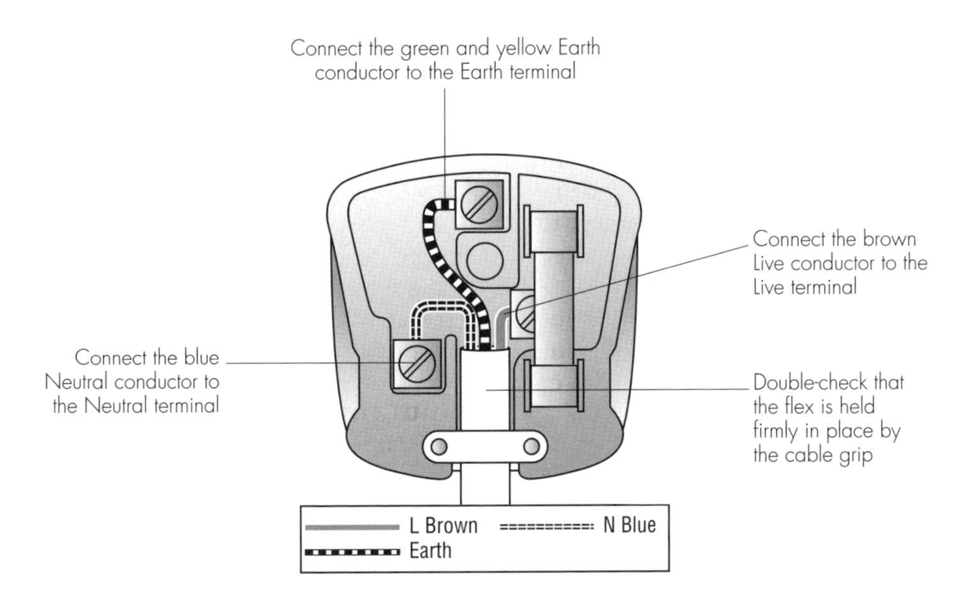

Wiring a plug: Step 2.

Before you start the wiring, check whether the plug cover needs to be fed over the flex first. There's nothing more infuriating than finding this out when you have finished wiring and having to start again.

Twist the fine strands that make up each core with your fingers so that they don't splay out when you try to put them in the terminal hole - two or three twists will do the job. First put the brown live wire into its terminal hole, making sure that the bare copper passes right through the hole. Then tighten the screw with the 4 mm screwdriver; it may be easier to remove the fuse and the live terminal from the plug first, then secure the live wire in the terminal before replacing it in the plug. Next fit the blue neutral wire into its terminal and screw it down tightly (again it may be easier to do this by removing the whole pin first). The green+yellow earth core goes to the earth pin, which is always a bit longer than the other two pins. As with the others, it may be easier to remove the pin from the plug in order to make the connection.

Double insulated appliances, such as plastic encased radios and vacuum cleaners, are wired up with twin-core flex, which has no green+yellow earth core in it. This is fine because the appliance has no metal parts that you can come in contact with. Such appliances should have a label with a symbol that looks like a small square with another square drawn inside it. If the appliance you are dealing with does in fact have metal parts that can be touched but is nevertheless wired with twin-core flex, it may be that someone has replaced the original twin-core and earth flex. In this case, make sure the appliance is unplugged and have a quick check to see if there is an earth terminal in the appliance.

Put the cord grip back over the flex, gripping the outer insulation firmly, and fit the two screws. On plugs with plastic grippers push the flex into the grip with the square nose of the pliers. Give the flex a good tug to make sure that it cannot come out of the cord grip. Replace the plug cover, which must fit evenly with no gaps around the edges. If there is a gap, have a look inside the plug to make sure the three wires are lying neatly next to each other and are all in the bottom of the plug, not over a plastic edge or ridge. Fit the screw and tighten it up, although not too hard or you will crack the plastic cover. There should be no coloured cores visible outside the plug. If any of the plugs in your house are like this, you must rewire them straight away as the only thing separating you from 230 volts of electricity is a millimetre of coloured soft plastic.

Mounting cable on the surface

Before you start

Although most people prefer a streamlined look and do not want to have visible surface-mounted cables inside the house, this is by far the easiest method of running power cables in workshops and garages.

Tools and materials

There are various types and sizes of plastic cable clip, so make sure the clips match the cable you are using. The cable size is stamped in the same lettering on the top of cable clips, for example, a clip with "2.5" stamped on it should be used for 2.5 mm^2 cable. The clips themselves come in several colours - white, grey and black. Most clips have special masonry nails so that they can be hammered into brick, concrete, timber and cement mortar. If you lose a masonry nail from a clip, don't try and replace it with an ordinary timber nail, as it will only bend over immediately - just use a new clip. There are also larger, black, D-shaped clips specially designed to suit armoured cable fixed to outside walls.

If there is any chance of damage from machinery, bikes or cars, the cable should be surface mounted in trunking (see pages 60-62). Galvanized steel trunking and junction boxes are recommended in these circumstances.

You will also need the following tools and materials: tape measure, pencil, claw hammer, safety goggles, long-nosed pliers and for fixing clips for armoured cable, electric drill, drill bit and wall plugs suited to the wall or ceiling construction, round-head screws, screwdriver.

Method

Once a two-way consumer unit has been installed on the workshop or garage wall, surface-mounted cabling can be run to the sockets and lights. Before you start clipping the cable in position, it is important to remove any twists and bends from the cable so that it looks neat and tidy on the wall. Never wire up both ends of the cable before fixing it into the wall, because then it will be impossible to straighten it out.

When you're sure that the cable will run smoothly and follow the route, fix the first clip in place. Position the clip so that the nail is under the cable; this is done so that even if the clip becomes loose the cable will not fall out. With the nail fully up in the clip body, snap the clip over the cable and hold the clip in place with the tip of one finger to stop it falling off. Strike the nail squarely with the hammer to get it started. If it is driving well then continue with firm blows until the nail is fully home. On a safety note, beware that masonry nails are made from high tensile steel and can shatter if they are hit with a glancing blow from a hammer. Always wear safety goggles when driving in masonry nails. If the nail starts going off at a strange angle, straighten it up before continuing otherwise it will probably chip out a chunk of masonry. A top tip for beginners is to hold the nail with a pair of long-nosed pliers so that you don't hit your finger!

Once the first clip is in place the difficult part is over, smooth the cable on to the wall taking care to remove any kinks or wrinkles. Keeping tension on the cable so that it lies flat, fix the next clip in place leaving a 100 mm gap from the first one. Remember to keep the nail under the cable. Using

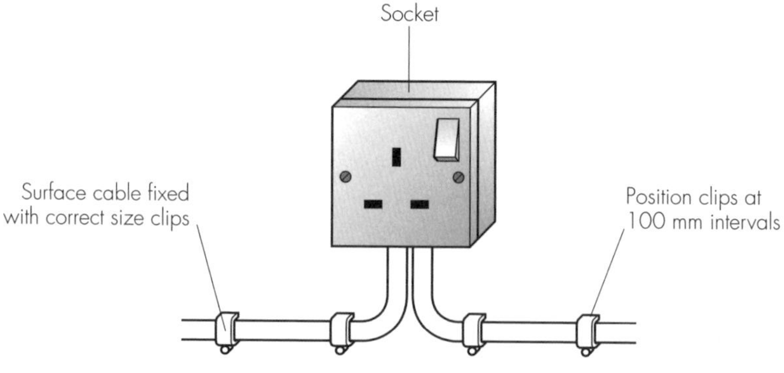

Fixing cable to a wall using clips.

this technique, continue along the cable run placing clips every 100 mm until you get to a corner. Keep the clips looking neat and tidy - nothing looks worse than oddly-spaced cable clips hammered in at strange angles. Immediately before the bend, fix the cable with a clip, smooth the cable round the corner so that it lies flat against the wall, and fix it with another clip before continuing along the next straight run. Although the cable is oval in section it can still be bent around a flat curve with a little care. Never use pliers or any other sharp tool to bend the cable since this can damage the outer insulation.

To surface mount armoured cable, you need to use the larger, D-shaped cable clips. Just as when using ordinary clips, the spacing of the clips should be regular and the cable should be level and smooth. Carefully mark all the positions for the clips, at 300 mm intervals, using a long piece of wood and the tape measure. Drill a hole at the first mark using a suitable masonry bit, and tap in a wall plug. Open up the cable clip and fit it around the armoured cable. Using a round-head screw and the screwdriver, fix the clip to the wall, keeping the screw under the cable. If the cable run is outside it is best to use galvanized, or failing that zinc-plated screws, since these won't leave rust marks on your wall.

Mounting cable in trunking

Before you start

Cable trunking is a quick and easy way of running cables to add lighting or power circuits in your house without major damage to the decorations. It is impossible to fit buried ducted cables without major repairs to plaster, paint and wallpaper afterwards.

Tools and materials

Basic interior trunking is usually made of white plastic and is rectangular in section with a clip-on lid that makes installing the cable in the ducting very easy. It has a self-adhesive backing, which makes it quick to fit. It is now easily available from DIY stores, often cut into lengths to fit easily into a car. Trunking is available in several sizes, so make a plan of the wiring that you will be installing in the trunk so that you know how much of which size to buy for each part of the system. For example, 1.00 mm^2 cable running from the ceiling to a light switch will need the smallest size of trunking, whereas a double-pole ceiling switch will require the largest trunking because you may need to put two large 4 mm^2 cables in it.

There are also a number of proprietary corners, angles, bends and blank ends which clip easily to the plastic trunking to make a smooth, professional-looking job.

There is another type of trunking, which is tubular and made in plastic or galvanized steel. It is used in conjunction with small, round junction boxes. This tubular system is not normally fitted indoors as it looks rather industrial, but it is ideal for the garage or workshop because it is strong and lasts well. The steel trunking is quite difficult to fit because all the wires must be pulled through the tubes using a thin wire or nylon cord called a mouse as the trunking is being assembled. You will also need to use a pipe-threading machine, which can make this system rather expensive, although you can hire the machine from a good tool hire shop.

You will also need the following tools: pencil, hacksaw with a fine blade, electric drill, masonary drill bit and, for galvanized steel tubular trunking, pipe-threading machine.

Method

The cable will probably enter the room from the loft, a built-in cupboard or a corner. Making sure the cable is safely out of the way, drill a small hole at this point, tight in the corner. In this way, the cable will immediately enter the trunking.

Offer the trunking into position and mark one end if it needs to be cut at a particular angle or shape to fit neatly. Cut the trunking using the hacksaw. Always cut the end with the angle or shape first, so that if you make a

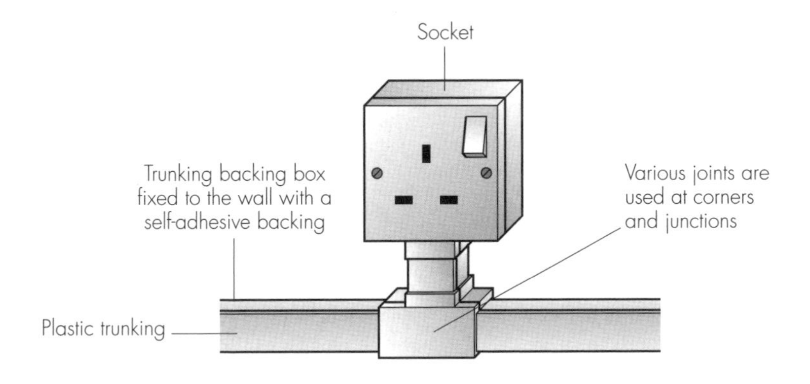

Mounting cable in plastic trunking.

mistake there's always another chance to rectify the situation before you cut the other end. Offer the trunking into position again to check the first cut and if all is well, mark and cut the trunking to its final length.

Take the clip-on lid off the length of trunking. Peel the paper backing off the self-adhesive strip and press the trunking firmly into position. If it does not have a really good hold, drill a few holes through the trunking, tap in wall plugs and secure the trunking with screws.

Use proprietary corners and joints to join one straight run of trunking to another. These normally push on to the ends of the main trunking. If necessary, a screw and wall plug can also be used to secure a corner or other joint in place.

On the last run before a fitting, check and double-check the length of trunking required and the shape the end needs to be. When you are sure, cut and fit the last length of trunking.

All tubular trunking systems work in a similar way. Secure the small, round junction boxes in place with wall plugs and screws. Cut the straight lengths of tubing to the correct length with a hacksaw and fit them between the junction boxes. The plastic trunking has small glue-on ends so that it can be cut to fit into each junction box and there is a nut in the box to secure the end of the tube. Use screws and wall plugs to secure the lengths of trunking at regular intervals with the accompanying heavy-duty cable clips.

Running cable under floorboards

Before you start

Although modern houses usually have a solid concrete floor at ground level, nearly all houses have wooden floors on every other level. Older houses also have timber floors on joists at ground level. Unfortunately modern building trends mean that wooden floors made of solid timber planks are rarely viable because they are too expensive. Since 1970 most houses have floors made of chipboard panels. These have tongue-and-groove edges, which lock the panels in place and can be very difficult to lift for laying cable in the space between the floor and the ceiling below. Traditional timber plank floors are much easier to work on because individual boards can be lifted and cables laid underneath.

Tools and materials

You will need the following tools: a pry bar, a claw hammer, an electric drill, a short spade bit or joist brace, a rasp, a cross-head screwdriver, some 50 mm (2 in) cross-head screws.

A joist brace is a traditional tool that is specifically designed to drill holes in joists, although it can be quite expensive to buy.

Method

Referring to your wiring plan, roll back the carpet in the room where you wish to lay cables under the floor. Mark any boards you want to raise and the adjacent ones with sticky tape so that you know which way round they were. In this way, you can ensure they

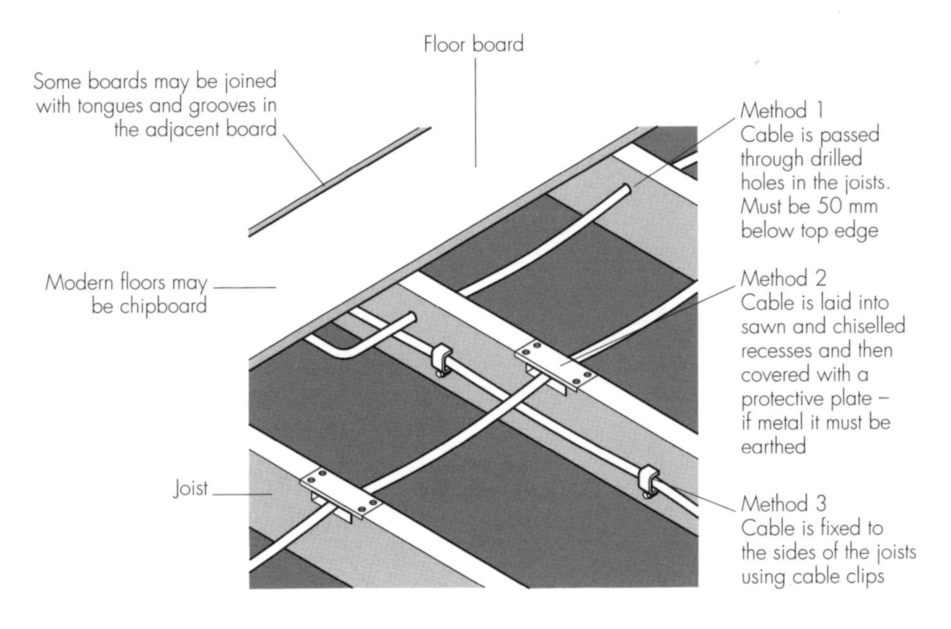

Laying cables under a wooden floor: three options.

fit back snugly when you replace them and you do not accidentally drive a nail through a pipe or cable.

Jam the pry bar in between the ends of two boards to lever the end of the board you want to raise. You will find that it usually requires a lot of force to lift the board as the nails will probably have been there for years and will be well set into the joists. When it is removed, turn the board over and tap the nails back through the boards, before pulling them out with the claw hammer. This is an essential safety precaution as it's very easy to step backwards on to a protruding nail and end up in the local casualty department for the best part of a day!

Use the drill and spade bit or the joist brace to drill holes through the centre of the side face of the joists through which you want the cable to pass. If there is already a hole at that particular point in the joist, do not drill another hole as it could weaken the timber. Either fiddle the cable through the existing hole or, if necessary, enlarge it slightly with a rasp. If it is impossible to enlarge the hole, drill another one further along the joist and reroute the cable slightly to avoid creating a weak point in the timber.

Thread the cable through the holes in the joists according to your wiring plan. Then replace the floorboards using cross-head screws, instead of floorboard nails, because they grip much more effectively and are an excellent way of avoiding creaking, squeaking floorboards.

Chipboard floors are more difficult to lift as the tongue-and-groove joints between the panels hold all the panels together as one sheet. They are not easy to deal with without damage, so make a note of the thickness and dimensions of the panel and the design of the tongue-and-groove, so you can check that the DIY store stocks panels to match.

First remove all the cross-head screws that usually hold these panels down. Use the pry bar or a big screwdriver to carefully prise up one panel. There will nearly always be some damage to the panels, so try to limit it to just this one panel. You could minimize the risk if you can find the last panel that was originally laid in the room and work from that direction. As an alternative, it may be possible to remove the first panel by breaking off the tongue around the edge. If you take this option, screw blocks of wood to the nearest joist underneath the floor to act as a platform on which to put the butting edges of chipboard. You must position the timber platform at the correct height to rest the panels in their original alignment, otherwise they will at best creak and at worst break. Once you have removed the first panel, any others should come out more easily. Chipboard sheets are not expensive so don't be afraid to get stuck in and wreck one – it can easily be replaced.

Running cable in solid walls

Before you start

Nearly all houses have cables fitted beneath the surface of the wall. This is the neatest and tidiest method when all is finished, but without doubt it is the messiest while it is being done!

All walls that have been chased for buried cables will require, at the very least, to be replastered and repainted, and usually the whole room will need redecorating so that everything matches. Wallpaper is always impossible to repair even if you do still have some of the original wallpaper, so you might as well strip all the paper off with a steam stripper before you start the cabling work.

Tools and materials

There are quite a few fancy electric tools for chasing the cable runs in solid walls, most of these involve an impact-driven chisel, which has a swan-shaped neck. This chasing equipment is quite expensive, but if you already have an SDS chuck and a rotary stop hammer drill, you can buy the swan-neck chisel quite cheaply. Ordinary mortals must make do with a hammer and chisel, which are perfectly good for the job!

All cables buried in walls must be in ducting. If they are not and are simply plastered into the wall, it will be impossible to pull out and replace the cables at a later date should the need arise. Ducting comes in various sizes to suit different cable, so refer to your wiring plan and the different sizes of cable needed for the various parts of the system and make a list of the lengths and diameters of the ducts you will require. Proprietary push-on corners and bends are also available.

You will also need the following tools and materials: permanent marker pen, electric drill, masonry bit, hammer, chisel, safety goggles, gloves, thin string or wire mouse, instant contact adhesive, insulation tape. You will also need some sand and cement mix, and plaster, and to complete the job later, general decorating, painting and papering equipment.

Method

Referring to your wiring plan, mark the cable runs on to the wall with permanent marker pen. Using the electric drill and masonry bit, drill a long succession of closely spaced holes along the pen lines. This may seem tedious, but it will ensure that you eventually produce a nice square-sided channel that is easy to repair and won't show afterwards. Each hole should be about 30 mm deep and should nearly touch the edge of the last hole. The channel in the wall will need to be roughly 30 mm wide, but the exact width will depend on the ducting you wish to use. The ducting should sit about 20 mm below the original surface to allow for making good the wall, so make sure you allow for this in the overall depth. Once you have drilled the rows of holes, use the chisel and hammer to knock out a channel with a nice flat bottom. Make sure, at the very least, that you wear the protective gloves and goggles for this operation.

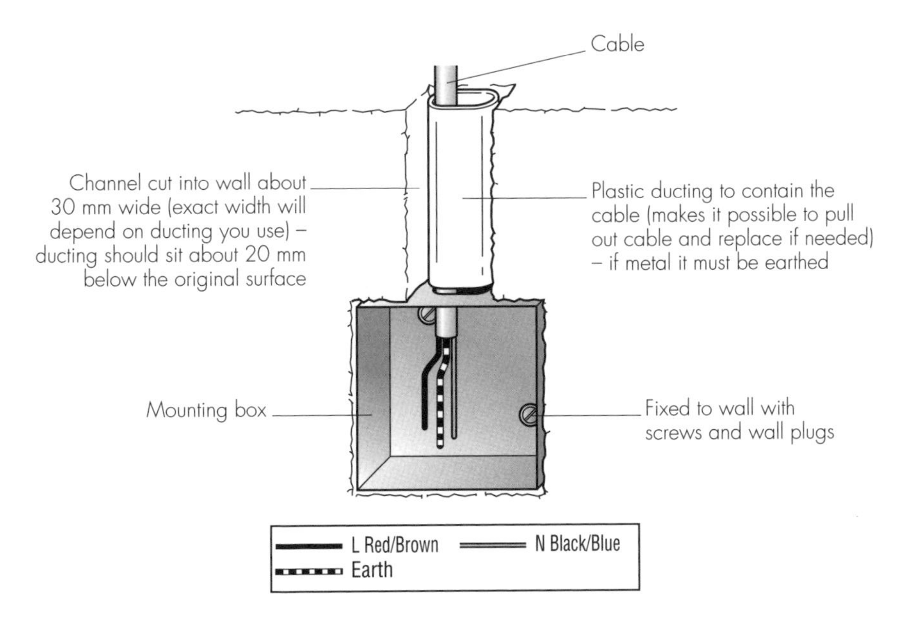

Installing cable in solid walls.

Now take the plastic ducting and thread a thin string through it ready to use as a mouse (that is, for pulling the cable through) later. Tie the ends of the string to a couple of spoons or something similar so that they won't disappear back into the ducts. Apply the adhesive to the ducting and press it firmly into the channel.

Fill in the hole over the ducting to within 4 mm of the surrounding surface. When the mortar has gone off, skim the remaining 4 mm with plaster. Plastering is a specialist skill and you may have to allow for the extra cost of employing a plasterer if you want a seamless finish to the job.

When you are ready to run cable through the ducting attach the cable to the string or wire mouse. Join them securely by laying the two ends side by side for about 200 mm and taping them tightly together with insulation tape. Pull the cable through by pulling firmly, but not hard, on the mouse. "Wiggle and fiddle" is best!

Running cable in hollow walls

Before you start

The non-load-bearing walls in modern houses are normally constructed using plasterboard on each side of a timber framework. This leaves a convenient

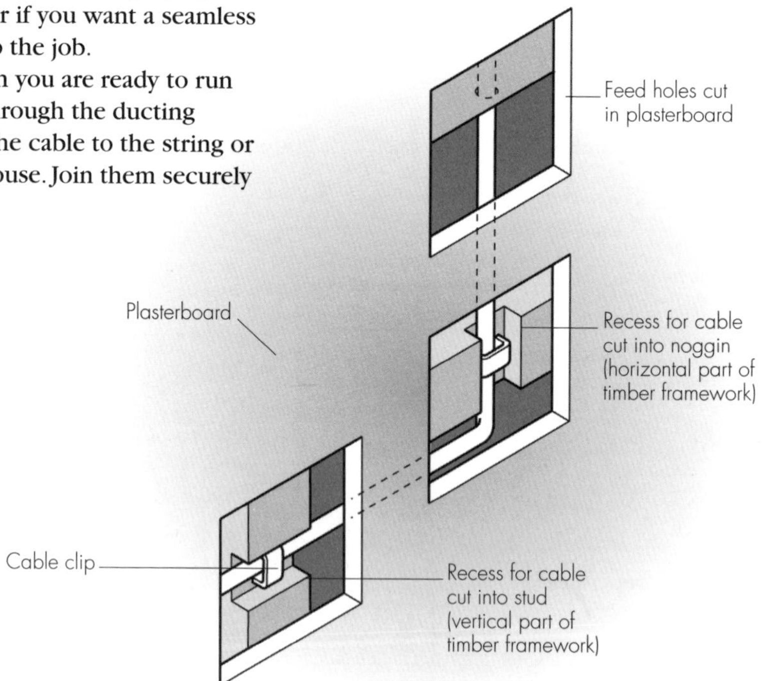

Installing cable in hollow walls.

hollow space for running cables. You can usually detect this type of wall construction by tapping it with a screwdriver handle - it will sound quite hollow. If you tap along a long wall you will hear the tone change when you pass over the timber framework behind and you may even see the small dents left by the plasterboard nails in the surface. The cabling cannot be run in the same position as the timber, so you may have to move the planned socket or switch a little to the left or right to accommodate this.

Running cable in hollow walls is fairly simple. However, it does require a lot of patience to fish the cable through from the cavity as invariably the new cable won't hang exactly behind the hole in the plasterboard so that you can see it.

Tools and materials

You will need the following tools and materials: electric drill, jigsaw, 25 mm spade drill bit, 60 mm diameter hole-cutting drill tool, plumb bob, a small nut, string, 1 m of fishing line or thin wire, hammer, wood chisel. A hole-cutting drill tool is sometimes called a trepanning tool because it cuts a perfect disc. The wire can easily be made from a thin wire coat hanger.

Method

Referring to your cable plan, work out where the cables will run on site. They will usually run through the ceiling space or under the floor and then down into the space in a hollow wall. You will need to lift the floorboards in the room above the wall cavity, so measure carefully to identify the position of the wall beneath. Then, using the spade bit, drill a large hole through any ceiling plasterboard and the horizontal wooden timber forming the top part of the partition wall's framework. The wall cavity will then be seen.

Go back to the room where you want the electrical fitting and, measuring carefully and using a plumb bob, mark its position on the wall in a direct vertical line below the hole in the ceiling space. Use a jigsaw or a hole-cutting drill tool to cut a 60 mm hole through the plasterboard.

Back on the floor above, tie a small nut on to a piece of string and dangle the string down through the hole in the ceiling. Lower the nut into the cavity. If it encounters a horizontal timber strut (part of the framework, and known as a 'noggin') on its way, cut an opening in the plasterboard and chisel out a channel to allow it to pass (refer to the illustration on page 66). Tie the string off securely on something in the ceiling space. Make a small hook in the end of the length of wire. Poke it through the hole in the wall and try to catch the dangling string in the cavity. You will probably need to bend the wire in order to get it into all the corners of the cavity space before you hook the string. When you have hooked the string, pull it and the nut out through the hole into the room.

Go back to the floor above and attach the cable to the other end of the string in the ceiling space. Do this by laying the end of the string along the end of the cable and binding the two together with insulation tape so that there are no loose ends or bulky shapes to catch on anything as the cable finds its way down the wall. From the room below, pull the string to bring the cable down through the cavity and the hole in the wall.

Pulling cables through cavity wall spaces is a fiddly job and can feel impossible when working by yourself. To make it easier, enlist the help of someone else who is willing to help feed the cable or string down into the wall while you pull it through.

Fitting ceiling roses and pull-cord switches

Before you start

Fitting ceiling roses and switches is a little different to fitting other electrical components on the wall or ceiling, because they need to withstand considerable weight. Pull-cords get yanked hard and ceiling roses need to support the weight of light fittings.

Most ceilings are made of plasterboard or plaster. They are therefore fragile and easily damaged. However, if your house has an open beam ceiling, the switches or ceiling roses can be screwed directly into the beams.

Tools and materials

Most ceiling roses and switches have two slotted holes in the base of the fitting, which are usually about 5 mm in diameter, to take the supporting screws. These slotted holes allow the fitting to be moved slightly in the event that the holes in the ceiling do not quite line up.

There is an exotic range of expanding screws and bolts for plaster and plasterboard ceilings. They all work in the same way. The whole expanding part of the bolt goes into the ceiling. When the screw is tightened, the head flattens and covers a large surface area above the ceiling. They all have the advantage of being able to support a substantial amount of weight with relatively small screws. Once expanding fittings are in place they are impossible to remove, so if you lose the fitting in the ceiling space you must use a completely new one.

There is also a more sophisticated fitting available, made especially for use with plasterboard. These fittings look like a large metal corkscrew with a smaller self-tapping screw down the centre. They are very clever because they do not require a predrilled hole. You simply screw the aluminium body directly into the plasterboard ceiling with a large cross-head screwdriver. In this way, the sharp point on the fitting drills and screws its way through the plasterboard. Self-tapping screws are used to secure them in position. However, this type of fitting does not work in lath and plaster ceilings such as you would find in a Victorian house, and the regular expanding fittings are a much better option in those circumstances.

You will also need the following tools and materials: an electric drill, twist and spade drill bits, a pencil, some plasterboard ceiling plugs, some screws, a screwdriver.

Method

Before drilling any holes in the ceiling make sure there are no power cables in the ceiling space anywhere near the site of the proposed ceiling rose or light switch.

Referring to your wiring plan, check and mark where the cable will come through the ceiling. Drill a small pilot hole through the mark to make sure that the cable can pass through the ceiling without a problem. Then run the cable to the hole and wire it into the fitting (see page 74).

Thread the cable through the large hole in the base of the ceiling rose or switch and place the base in its final position, making sure that the cable is not bent or kinked. Mark the centre of each hole position in pencil on the ceiling. Measure the head size of the expanding screw and select a drill bit that is about 1 mm bigger. You should also be able to check the correct drill size on a table on the packaging for the expanding screws. Drill the holes through the plasterboard and push the expanding screws into the plasterboard. Place the base of the rose or switch in position. Tighten the screws enough to expand their heads but not so much as to cause the plastic to crack.

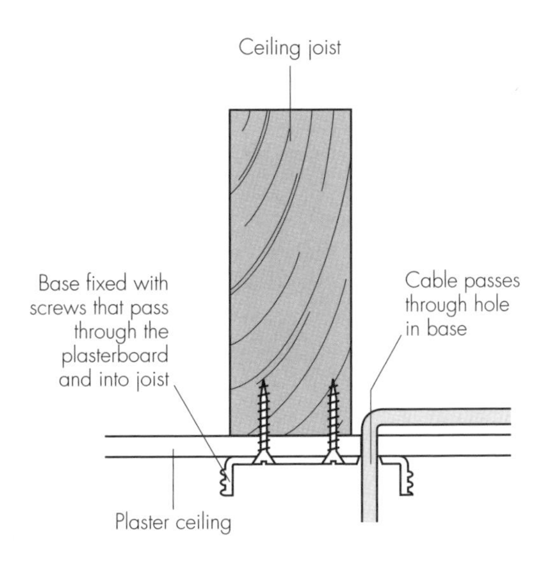

Base of a ceiling rose or switch fixed to a ceiling joist.

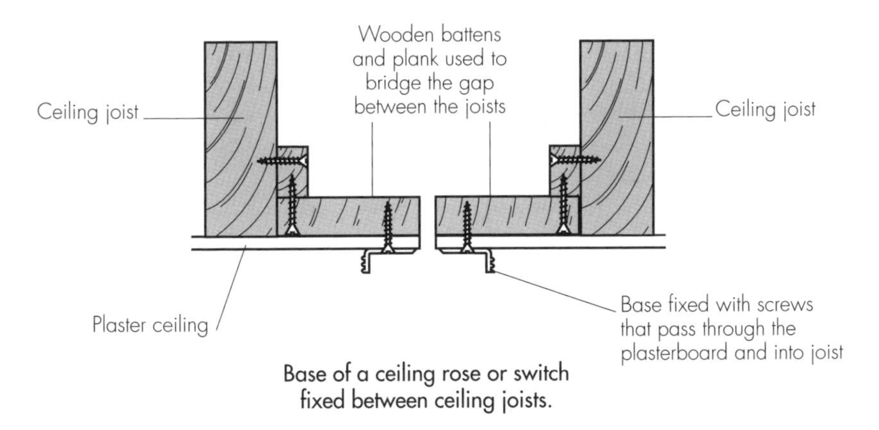

Base of a ceiling rose or switch fixed between ceiling joists.

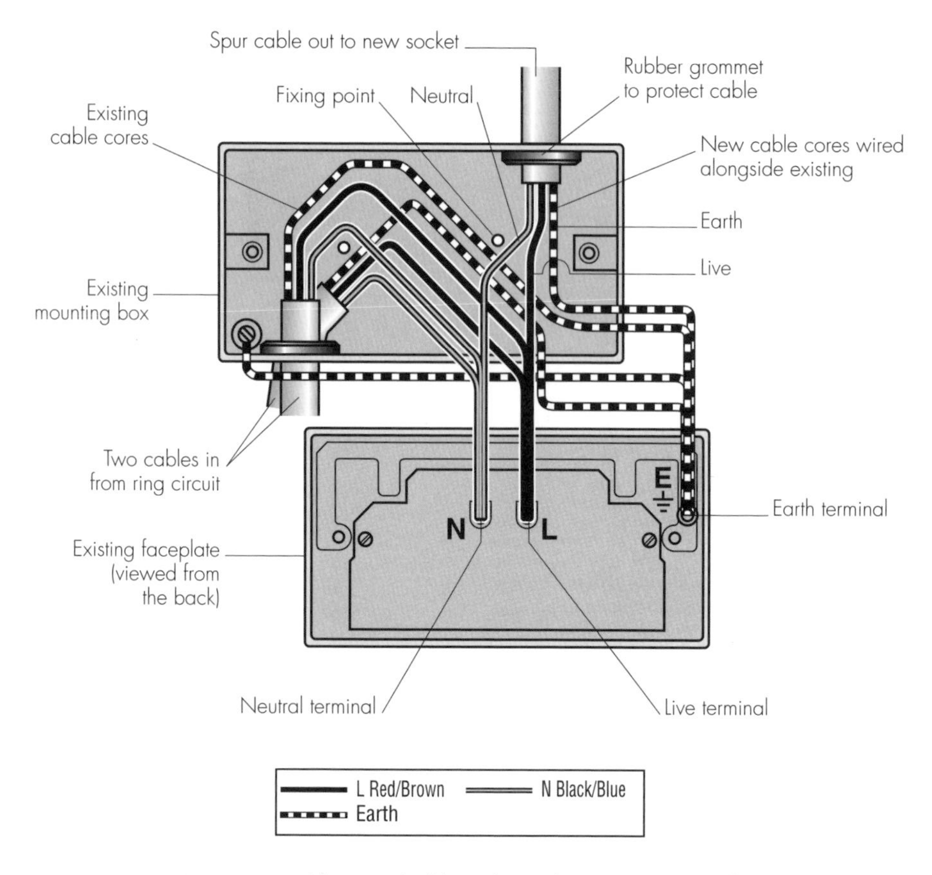

Connecting cables to a double socket with a spur running off.

Connecting cable to sockets

Before you start

It is important to be able to make proper wiring connections behind sockets, because they are on a ring circuit that must be continuous and one bad connection can overload the system. In a ring circuit the cable runs in a circle from the consumer unit or fuse box around each socket and then back to the consumer unit again. Although a badly wired socket may still work, there will be a break in the ring circuit, putting a higher load on parts of the circuit. Badly wired sockets may also fizz and crackle when a plug is inserted.

Tools and materials

Buy a good brand of socket. Not only will it be more robust, but the terminal layout is specially designed to be easily wired by electricians.

You will also need the following tools and materials: pliers, side cutters,

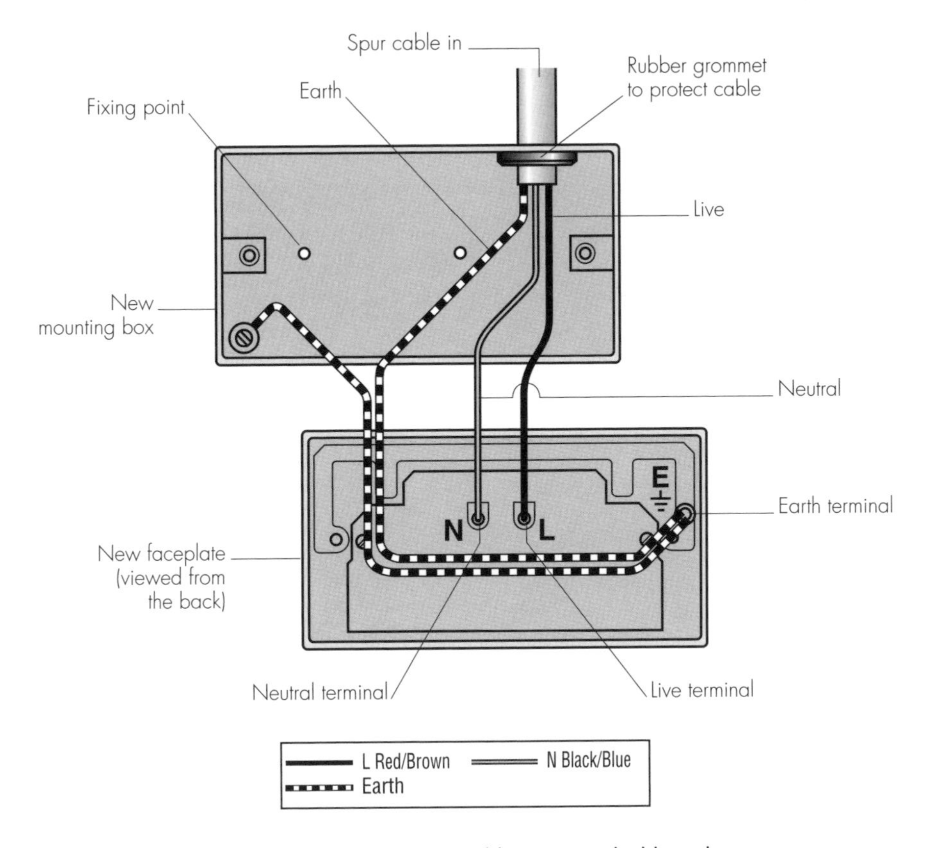

Connecting a spur extension cable to a new double socket.

wire strippers, green+yellow sleeving, 4 mm electrician's screwdriver, flat-head screwdriver.

Method

Switch off the circuit at the MCB (remove the fuse) and check that the circuit is DEAD. There will be two cables coming through the backing box on the wall if the new socket is on a ring circuit, one coming from the previous socket and one going to the next socket on the circuit. If the new socket is at the end of a radial circuit or a spur, there will only be one cable. In order to make wiring the socket easy, you will need to have about 200 mm of cable protruding from the wall. You might get away with 100 mm, but it will be more of a struggle.

Using the pliers to grip the bare earth core, strip 100 mm of the outer insulation away from the cable. Cut away the outer insulation with the pliers or side cutters. Use the wire strippers to strip 20 mm of insulation off the red/brown live and black/blue neutral cores. Fold the bare copper ends over in half and nip them tight with the pliers, which stops them from being pulled out of the screw terminals easily. Cut 80 mm length(s) of green+yellow sleeving and slip it over the bare earth core(s). Then fold over the end of the core(s) and nip flat.

Using the electrician's screwdriver, loosen all three terminals screws on the back of the socket so that you can no longer see any of the screws visible in the terminal holes and you will be able to insert all the cores easily. Connect the green+yellow earth core(s) under the terminal marked "E", the red/brown live core(s) under the terminal marked "L" and the black/blue core(s) under the terminal marked "N". Make sure that all the cores are pushed fully home into the terminal holes and you can see them out of the other side before tightening the screws. There should not be very much, if any, copper core visible. Give the cable a good pull to make sure that the cores do not pull out of the terminals. Secure the socket cover with the two screws provided.

Connecting cable to switches

Before you start

There are two main types of switch that you may want to connect – ordinary wallplate lighting switches and heavy-duty double-pole switches for showers, immersion heaters and wall heaters. The pull-cord ceiling lighting switches used in bathrooms are wired up in exactly the same way as the wallplate switches. Wallplate switches are easy to wire up because 1.00 mm^2 or 1.5 mm^2 cable is used; the 4 mm^2 cable usually needed for double-pole switches is much more difficult to bend and work with.

Tools and materials

You will need the following tools: pliers, side cutters, wire strippers, green+yellow sleeving, red/brown insulation tape, 4 mm electrician's screwdriver, earth core, flat-head screwdriver.

Method for wallplate switches

If the new wallplate switch is a one-way switch, there will be one cable coming through the backing box on the wall from either the ceiling rose in a loop-in system or from a junction box. With a two-way switch, there will be a second cable connecting this switch to its partnering two-way switch. In order to make wiring the switch easy, you will need to have about 200 mm of cable protruding from the wall. You might get away with just 100 mm, but it will be more of a struggle.

Using the pliers to grip the bare earth core, strip 100 mm of the outer insulation away from the cable. Cut away the outer insulation with the pliers or side cutters. Use the wire strippers to strip 20 mm of insulation off the red/brown live and black/blue neutral cores. Fold the bare copper ends over in half and nip them tight with the pliers, which stops them from being pulled out of the screw terminals easily. Cut 80 mm length(s) of green+yellow sleeving and slip it over the bare earth core(s). Then fold over the end of the core(s) and nip flat.

First use the electrician's screwdriver to connect the green/yellow earth core(s) to the earth terminal at the rear of the backing box. If the switch faceplate is made of metal, you will need to make a short earth strap to connect the backing box to the faceplate. Cut 100 mm of earth core, strip 20 mm of insulation off each end and fold the ends over in half. Now connect the earth strap to the backing box and the same earth terminal in the backing box that you have already used.

Wrap some red/brown insulation tape around the black/blue insulation on the neutral core(s), to indicate that it/they will become live when the switch is turned on. (Effectively the red/brown live and the black/blue neutral cores are live when the switch is on.) Connect either the red/brown or the black/blue to the terminal marked "COM" and the other to the terminal marked "L1" (it does not matter which way round).

Method for double-pole switches

Under BS 7671 bathrooms are zoned and only SELV should be used and only suitable heaters/equipment are permitted in certain zones. Double-pole switches for use in bathrooms and shower rooms are designed to switch off both the neutral and the live wires to totally isolate the appliance. If you look at the

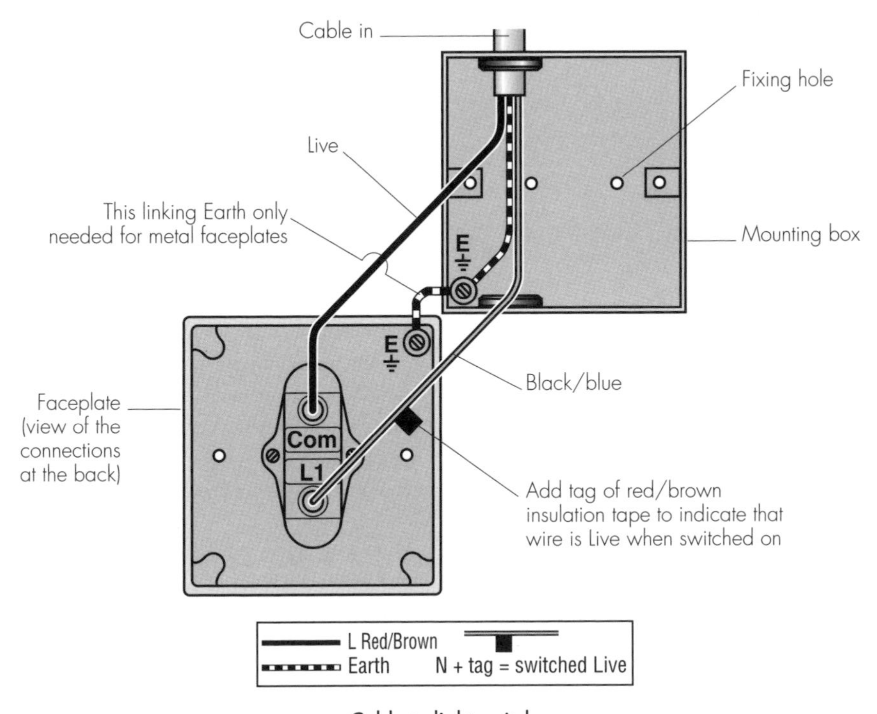

Cable to light switch.

connections on the back of a double-pole switch, you will see they are labelled "LOAD L", "LOAD N", "SUPPLY L" AND "SUPPLY N". There will also be one or two earth terminals depending on the model of switch.

There will be two cables coming through the backing box, one coming in from the consumer unit or fuse box and the other going to the appliance. Strip and prepare the cable in the same way as for wallplate switches. Connect the red/brown live and black/blue neutral cores in the cable bringing power in under the "L" and "N" supply terminals respectively. Connect the cable going from the switch to the appliance under the "L" and "N" load terminals. Connect both the earth cores under the earth terminal on the switch. Make sure all the cores are secure by giving the cables a good pull before closing up the switch cover.

Connecting cable to ceiling roses

Before you start

Ceiling roses are fiddly and tricky to wire up as there are quite a few cables entering a relatively small fitting. If the lighting is on a loop-in circuit, there will be three cables entering the ceiling rose. One is the loop in from the consumer unit/fuse box or previous ceiling rose in the circuit and the second is the loop out to the next ceiling rose. The third cable goes to the light switch. If you are wiring up the last ceiling rose in a loop-in circuit there will only be two cables entering the rose with no loop out.

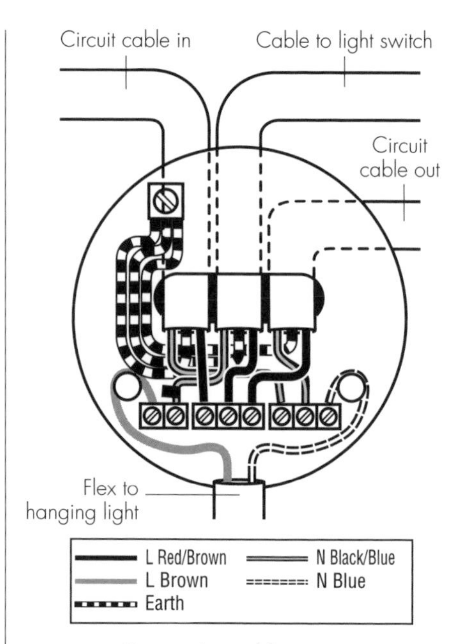

Connecting cable to rose.

Study the diagram carefully before you start because it is easy to wire up loop-in ceiling roses incorrectly, so that the lights are either permanently on, not working at all or constantly tripping out the MCB or fuse in the consumer unit or fuse box. There are four terminal blocks in a ceiling rose, one each for live, neutral, earth and the switch connection. Usually three of the blocks have three terminal holes and screws and the remaining block has two, each hole to take one core.

Tools and materials

You will need the following tools to wire a ceiling rose: ballpoint pen, pliers, side cutters, wire strippers, green+yellow sleeving, red/brown insulation tape, 4 mm electrician's screwdriver, earth core, flat-head screwdriver.

Method

Label each of the cables entering the ceiling rose with ballpoint pen to indicate where it comes from or where it is going to. Write "LOOP IN" on the cable coming from the previous ceiling rose in the circuit, "LOOP OUT" on the cable going to the next ceiling rose if there is one on the circuit, and "SWITCH" on the cable going to the switch.

Using the pliers to grip the bare earth core, strip 50 mm of the outer insulation away from each cable. Cut away the outer insulation with the pliers or side cutters. Use the wire strippers to strip 10 mm of insulation off the red/brown live and black/blue neutral cores. Wrap a piece of red/brown insulation tape around the black/blue neutral core in the cable that goes to the switch, to indicate that it will be live when the switch is turned on. Fold the bare copper ends over in half and nip them tight with the pliers. This will prevent them from being pulled out of the screw terminals too easily. Cut 40 mm lengths of green+yellow sleeving and slip it over the bare earth cores. Then fold over the ends of the bare earth cores and nip them flat.

Connect all the earth cores to the earth terminal in the ceiling rose. Connect the red/brown live cores from the loop in, the loop out if there is one, and the red/brown core that goes to the switch to the loop in terminal. Connect all the black/blue neutral wires, except the one with the red/brown insulation tape, to the loop out terminal. The black/blue live core with the red/brown insulation tape goes into the remaining empty terminal block, which normally has two screw terminals. This is the "switched live terminal", and is sometimes labelled "L".

Now it only remains to wire up the flex that connects the hanging light fitting. If the fitting is made of plastic you can use twin-core cable, but if it has any metal parts you must use twin-core and earth cable, which has an earth core in it. Strip about 60 mm of outer insulation from the end that will go to the ceiling rose and then strip 15 mm of insulation from the end of each core. Twist the fine cores tight together with your fingers and fold them over in half.

Slip the ceiling rose cover over the flex, making sure that it's the right way up to screw on to the ceiling rose. Connect the green+yellow earth core from the light fitting in the remaining hole in the earth terminal block in the ceiling rose. Connect the neutral blue core from the light fitting to the neutral loop out terminal in the same way, with all the black/blue wires. The live brown wire from the light fitting goes to the smaller terminal block occupied by the black/blue live core with the red/brown tape on it, which returns from the switch and makes the light come on.

Check that the screws are tight, then hook the blue and brown wires over the strain relief hooks. You should not be able to see the brown, blue or earth wires of the flex once the cover is screwed on. If you can, you will need to shorten the wires so that only the outer white covering of the flex can be seen. Slide the rose cover up the flex and screw it into position.

Earthing

Earthing is often neglected by householders because it doesn't actually affect whether appliances work or not and so it is easy to ignore an earthing fault. Unfortunately, in particular circumstances, an earthing fault can be very, very dangerous, resulting in electrocution or death.

If you look outside your house, you may find the top couple of centimetres of a metal rod driven into the ground somewhere next to the house wall with the earth wire connected to it, so that if there is an earth fault in your house, the electricity is conducted safely to ground. You must never disconnect this earth rod or remove it. If you see that the earth core has come out of the clamp, replace it immediately.

Inside the house, you will notice that any metal pipes for hot and cold water in the bathroom, shower room, toilet and kitchen have an earth clamp around them and they are all linked together. The earth wires and clamps that link the pipes are called earth bonding. This bonding is connected to the large earth terminal in or near your consumer unit or fuse box and from there a large earth cable goes to the rod outside the house, to the cold water pipes and to the gas pipes if you have them.

Virtually all appliances require an earth connection in a 3-pin plug, the only exception being appliances that are double insulated. Double insulated appliances are easily identified by the small symbol, which looks like two squares one inside the

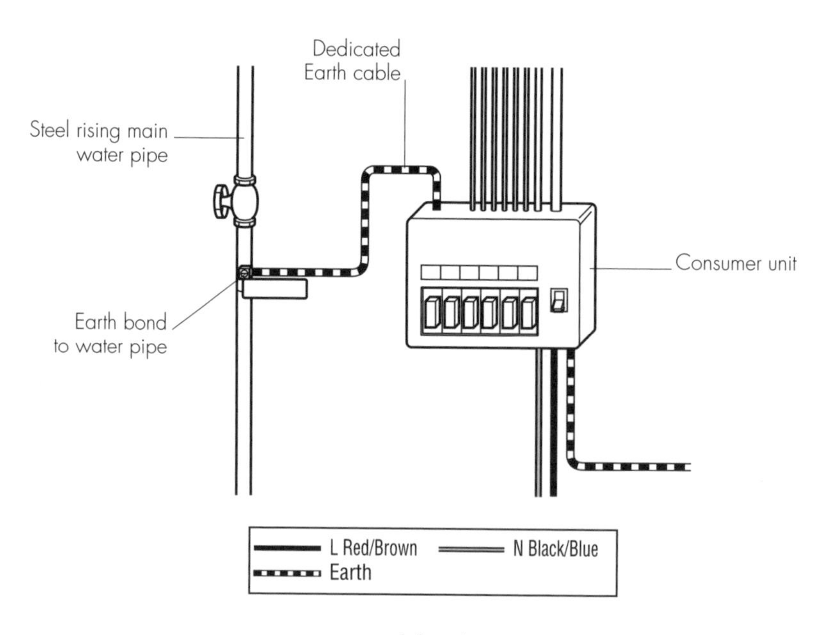

Earth bonding.

other, on their nameplate. They include appliances such as a plastic electric drill or radio. They have only a blue neutral and brown live core connected in the plug. All other appliances, especially those with metal casings or those that involve water, must be properly earthed. So when you do any work on a 3-pin plug, take care that the earth connection is secured properly before you begin to connect it.

The earth bonding of water pipes, cast iron baths, water heaters, metal sinks and central heating controls is most important because stray electrical currents can build up in these items and would otherwise have no safe way of dissipating. So for example, if you have refitted your bathroom, make sure that all the hot and cold metal pipes are connected with earth clamps to the nearest earth connection in a socket, fused connection unit (FCU) or switch.

Earth clamps comprise an aluminium band and screw system that is wrapped around the pipe and is tensioned with a spanner or screwdriver. Fit an earth clamp on each pipe and connect each clamp to the next with 4 mm^2 single-core PVC-sheathed earth cable. Connect the chain of earth clamps to the nearest earth terminal in a socket or FCU, so that they are connected back to the consumer unit or fuse box. Metal sinks and baths will have earth bolt points on the underside of them. Connect these into the chain of earth bonding with the same type of earth cable as for pipes. Occasionally metal pipes may have been repaired with plastic connectors. The plastic breaks the earth bonding in the water pipe. This situation is easy to remedy with two earth clamps and a length of 10 mm^2 earth cable. Fit one clamp on each side of the plastic joint and connect the two clamps with a length of earth cable.

Earthing is a most important part of electrical safety so it is a good idea to guarantee that it is effective. Get a professional electrician to test the earth bonding in your house and to issue you with a certificate. The certificate will come in handy if you ever sell the house or have to produce a new buyer's pack.

Replacing light switches

Before you start

Replacing light switches is a nice easy job because all the wiring will already be in place and all you have to do is change the switch faceplate. However there are several types of faceplate available, which operate single-way, two-way and intermediate switches. Different faceplates also have different numbers of switches - there may be one, two, three or four depending on how many lights the faceplate operates.

Tools and materials

To determine which type of switch faceplate you need, switch off the circuit at the MCB (remove the fuse) and check that the circuit is DEAD before inspecting the back of the existing faceplate. Remove the two cover screws and ease the faceplate away from the backing box; it may be stuck to the wallpaper or

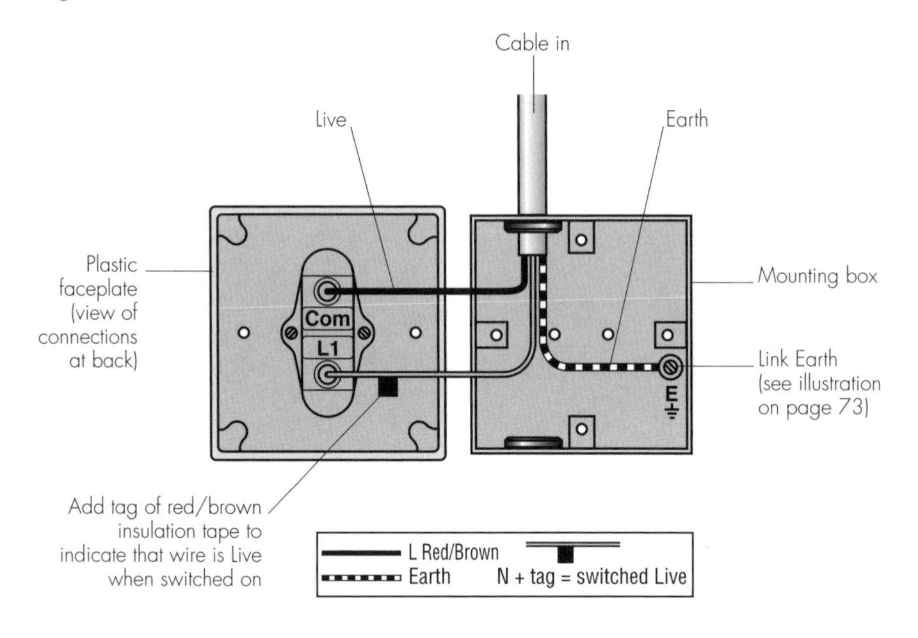

Replacing a single-way light switch.

paint, so you might need to prise it away lightly with a screwdriver. Using some coloured pens, make a diagram of the back of the faceplate on a sheet of paper.

If there are just two connections labelled "COM" and "L1", it is a single-way switch. Three connections labelled "COM", "L1" and "L2" indicate a two-way switch, probably operating a landing or hallway light. If there are four connections labelled "L1", "L2", "L3" and "L4", you have an intermediate switch, which is quite unusual. The intermediate switches are the third or fourth switches, which can be used to operate a light anywhere between two-way switches, for example along a long corridor with many doors opening on to it. Remember that there will be one set of wires from each of the switch buttons, or switch gangs. When counting the connections, don't count the earth connections because there are the same number of these in all switches. Check that your sketched diagram of the back of the light switch is accurate and then close up the cover.

When you order the switch faceplate, remember each switch button is called a gang, so a faceplate with three switches is called a 3-gang light switch. Don't worry if you can only get a two-way switch when you really only need a single-way switch, because it can still be used to do the job perfectly safely.

If you want to replace a light switch with a dimmer switch, you should note that you cannot normally buy a two-way dimmer switch and also that dimmer switches do not work on fluorescent lights, including the new economy bulbs, which have a fluorescent lamp inside them.

You will also need the following tools and materials: a flat-head screwdriver, some paper, some coloured pens or pencils, a roll of masking tape, a 4 mm electrician's screwdriver, some short lengths of earth core and earth sleeving and a roll of red/brown insulation tape.

Method

When you are ready to install the new light switch, turn off the circuit at the MCB (remove the fuse), check that the circuit is DEAD and open up the existing switch faceplate. Mark each wire with a small piece of masking tape and using your plan of the wiring, label each corresponding wire.

Using the electrician's screwdriver, remove all the wiring from the faceplate. If the new faceplate is a metal one, you will need to add a small earth strap from the faceplate to the backing box. This is simply a piece of bare earth wire covered in green+yellow earth sleeving that links from the earth terminal in the backing box to the earth terminal on the switch plate. If the old faceplate is plastic, there may not be an existing earth strap, but you must make one for a new metal faceplate.

If the new faceplate is the same type as the old one, refer to your wiring diagram and systematically rewire the rest of the connections on the new faceplate in the same configuration as they were before. If you are wiring a two-way switch on to a one-way circuit you will have two cables and three connections available, so connect the cores to COM and L1. Leave L2 empty.

Remember that the cable from the switch to the light fitting does not really have a neutral wire even though one of the cores is black/blue. When the switch is turned on, both the red/brown and black/blue cores can be live, so the black/blue should have a red/brown tag. If there isn't a tag, then you must wrap some red/brown insulation tape around the black/blue neutral yourself.

When you are satisfied that all the connections are secure, screw the faceplate securely in position on the backing box.

Replacing ceiling roses with different fittings

Before you start

Most houses have simple loop-in light fittings, which give you a single bulb hanging from the centre of your ceiling - these can be quite boring and not very conducive to setting a mood. If you take a trip to your local DIY store you will see very many different designs of ceiling light - some modern, some traditional and some futuristic. Most of these light fittings come complete right up to where they connect to the cables coming down through the ceiling. However, because the ceiling fittings are not usually big enough to incorporate the same connections that will go into a loop-in ceiling rose, you may have to make the connections above the ceiling in the loft or floor space in a 4-terminal junction box, as shown on page 80.

Tools and materials

You will need the following tools and materials: 4-terminal junction box or other connection box containing a 4-terminal connector, screwdriver, coloured pens or pencils, paper, masking tape, twin-core and earth cable, earth sleeving, off-cuts of timber, screws of various lengths, 13 mm hole saw and pliers.

Method

Switch off the circuit at the MCB (remove the fuse) and check that the circuit is DEAD. Make a diagram of the connections inside the existing loop-in ceiling rose. Use coloured pens to indicate each wire. Mark each wire with a small piece of masking tape and label the tape so that you know where each wire goes.

Refer to the diagrams on page 48 for instructions on wiring up the connection in your new 4-terminal junction box.

You must make sure that there is a continuous earth from the new light fitting all the way to the junction box. If the old light fitting is plastic, this might not have been the case and so it will not be on your original circuit diagram. Instead of a twin-core flex, which has a brown live and blue neutral core running to a pendant light, you will now need to use a piece of twin-core and earth cable to provide the connection from the 4-terminal junction box to the new light fitting. Strip and prepare the ends of the cable in the usual way, making sure that you fit earth

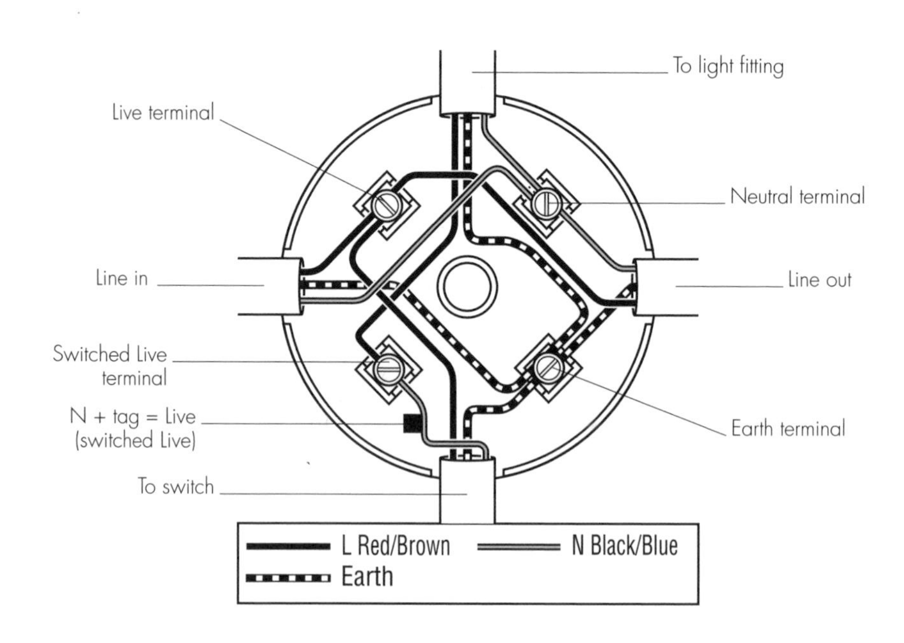

Junction box lighting connection.

sleeving on the bare ends of the earth core.

Identify the three terminal screw connectors in the new ceiling fitting and connect the three cores from the new cable to the fitting accordingly - red/brown to brown, black/blue to blue and earth to earth. Make sure that all the connections are completely secure.

Now fix the light fitting to the ceiling. If your ceiling is made of plasterboard, it may not be strong enough to support the full weight of the new light fitting. If this is the case, you may need to screw in an extra piece of timber between the two joists underneath the floorboards or in the loft space above the ceiling. Cut a length of timber to fit neatly between the two joists directly above where the light fitting will be connected and screw the piece of timber to the joists by means of blocks of wood. Screw the junction box on top of the timber, towards one side. Drill a 13 mm hole through the timber and pass the new cable through. If the new light fitting is in the same position as the old one, then there will already be a hole through the ceiling and you will simply need to drill through the timber. To secure the light fitting, use screws that are sufficiently long to pass through the new light fitting, the plasterboard and right into the strengthening timber. If you have doubts about the adequacy of screws, use long bolts with washers and nuts that will go on top of the strengthening timber.

Fitting a new ceiling rose

Before you start

Sometimes ceiling roses get damaged, perhaps when there is a wiring fault and the terminal block gets overheated or burnt, or when painters and decorators are at work. If part of the rose gets damaged, the best thing is to replace the whole unit. Although replacing the ceiling rose is quite an easy job, the sheer number of wires under the cover of a ceiling rose, most commonly on a loop-in circuit, can be off-putting.

The task necessitates turning off the main power supply, so you will either need to plan the work for daylight hours or run an extension lead and light from your next-door neighbour's house. It's important that you have good visibility to change a ceiling rose and see all those wires. Certainly you could just about use a torch, but only if you have a patient helper to hold the torch and to pass you the various tools and components you need.

Tools and materials

You will need the following tools: screwdriver, masking tape, coloured pens or pencils, screws, 4 mm electrician's screwdriver.

Method

Switch off the circuit at the MCB (remove the fuse) and check that the circuit is DEAD. Unscrew the ceiling rose cover. If you inspect the base plate of some models, you will see that there are four connection blocks, with the earth making the fourth connection under one large screw. Carefully label each wire with masking tape and draw

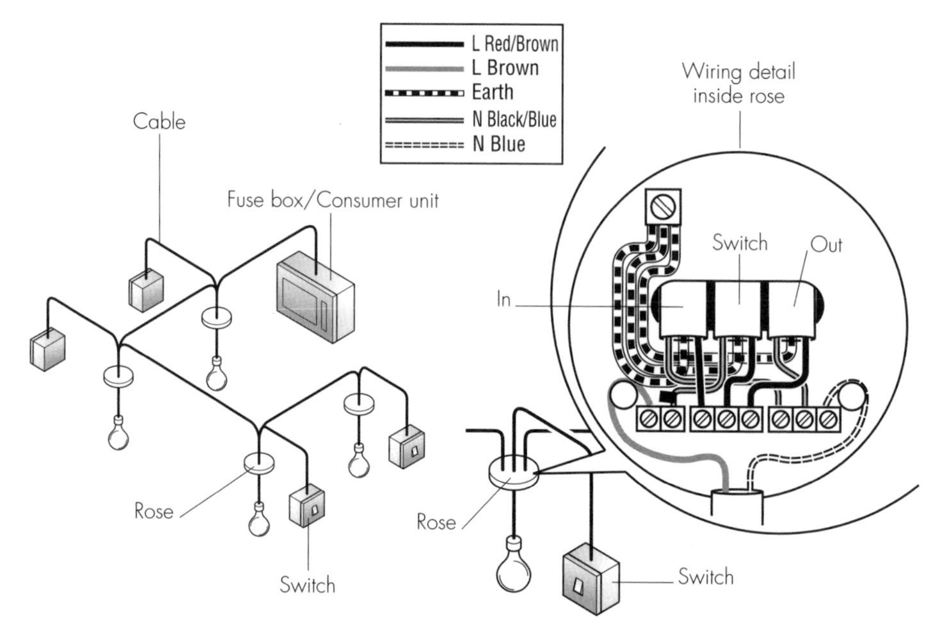

A loop-in ceiling rose system.

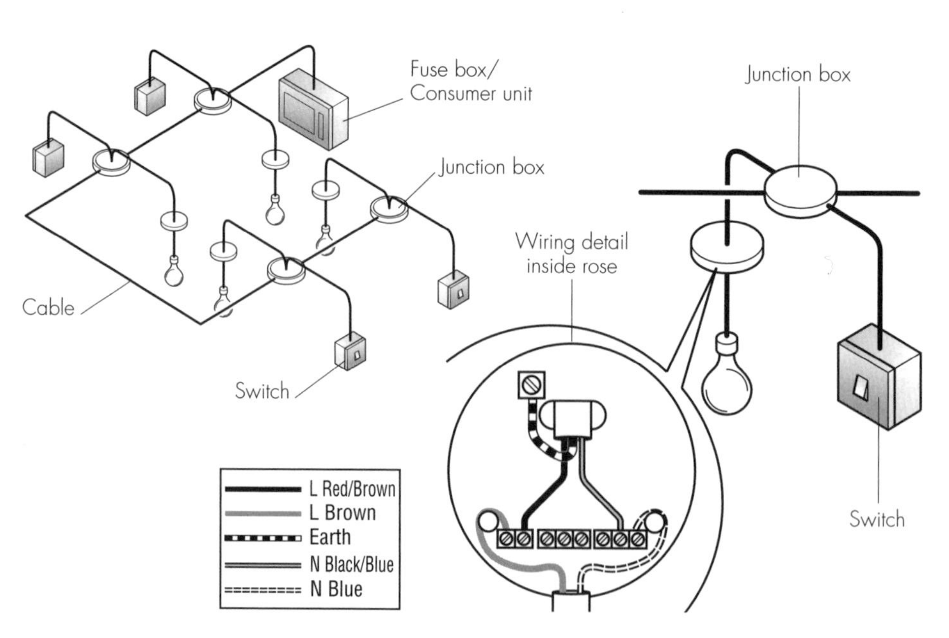

A junction box lighting system.

a clear diagram with coloured pens to show how the original unit is wired.

Unscrew the screws in the connection blocks and remove all the wires. Unscrew the old base plate from the ceiling and replace it with the new one, threading the cables through it. Secure the new backing box to the ceiling.

Now wire up the new ceiling rose in exactly the same way as the old one, with the wires going to the same four connection blocks as before. Remember to thread the flex from the light fitting through the new ceiling rose cover, which needs to be the right way up, before connecting it. Don't panic if the new ceiling rose is not exactly the same as the old one because it will still have four connections. When all the wires are securely connected, screw the ceiling rose cover in place.

You may find that the existing ceiling rose is a cosmetic one and does not contain four terminals like loop-in fittings. If this is the case, you are dealing with a lighting circuit on a junction box system (see page 80). This type of ceiling rose has only one incoming cable with three wires, the live, neutral and earth, and is usually connected into a small screw terminal block. Ceiling roses on a system like this are easy to replace because all you need to do is turn OFF the main power support, disconnect the cable from the terminal block, replace the old base plate with the new and connect the existing cable and the flex from the light fitting to the new ceiling rose. Always remember that red/brown goes to brown, black/blue goes to blue and, of course, earth goes to earth.

If the new light fitting is heavier than the previous one, you may need to reinforce the ceiling to take the extra weight. You can do this with a piece of timber secured with screws and wood blocks between the two joists, as described on page 81.

Fitting wall lights

Before you start

If you look in the DIY stores you will see there are literally hundreds of different types of wall light available in every conceivable design, shape and size. If you have chosen to have up-lighters where the bulb shines up from the top of the fitting, they need to be positioned at the correct height so that the bulb cannot be seen. Other wall lights should be sufficiently high that you can't hit them with your head when walking past.

As with all buried cable ducts the work is messy, because it will involve chasing ducting into the wall from the wall lights to the ceiling. You will need to redecorate afterwards, especially if you want a wallpaper finish. However, the effect will be worth the effort.

Tools and materials

You will need the following tools: a permanent marker pen, an electric drill, a masonry bit, a hammer, a chisel, safety goggles, gloves, some thin string, instant contact adhesive, a 4 mm electrician's screwdriver, wall plugs, screws, pliers, side cutters, wire strippers, lengths of 40 mm earth sleeving.

You will also need some ducting, 1.00 mm² twin-core and earth cable, some sand and cement mix, and plaster. In order to complete the job when you have finished the wiring, you will need general decorating, painting and papering equipment.

Method

First mark the location of the new light fittings on the wall. Make sure they will all be level if you are fitting more than one! Measure down from the ceiling to the position of the light fitting, as this will be the distance your eye will notice when the lights are on. Mark a cable run down the wall with the marker pen.

Using the electric drill and masonry bit, drill a long succession of closely spaced holes along the pen lines. This may seem tedious, but it will ensure that eventually you produce a nice square-sided channel that is easy to repair and won't show afterwards. Each hole should be about 30 mm deep and should nearly touch the edge of the last hole. The channel in the wall will need to be about 30 mm wide, but the exact width will depend on the ducting you wish to use. The ducting should sit about 20 mm below the original surface to allow for making good the wall, so make sure you allow for this in the overall depth. Once you have drilled the rows of

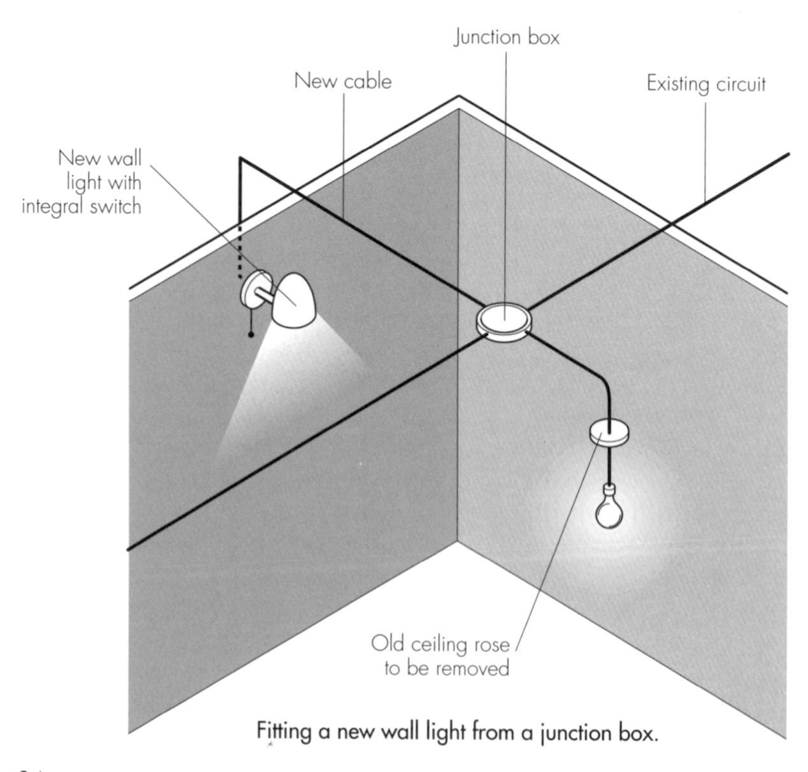

Fitting a new wall light from a junction box.

holes, use the chisel and hammer to knock out a channel with a nice flat bottom. Make sure that you wear the gloves and goggles at the very least for this operation.

Where the channel meets the ceiling, drill up with a long drill bit to access the space in the loft or under the floorboards above. Do some careful measurements and lift the floorboards in the room above to expose the newly-drilled hole.

Cut and fit the plastic ducting. Thread the string or wire mouse through it leaving plenty at each end of the ducting. Apply the adhesive to the ducting and press it firmly into the channel. Now run the new 1 mm^2 twin-core and earth cable from the junction box or loop-in fitting and down through the new ducting so that there is about 300 mm where the light will be situated. To run cable through the ducting, attach the cable to the string or wire mouse. Join them securely by laying the two ends side by side for about 200 mm and taping them tightly together with insulation tape. Pull the cable through by pulling firmly, but not hard, on the mouse.

Fill in over the ducting to within 4 mm of the surrounding surface. When the mortar has "gone off" (hardened), skim the remaining 4 mm with plaster. Plastering is a specialist skill and you may have to allow for the

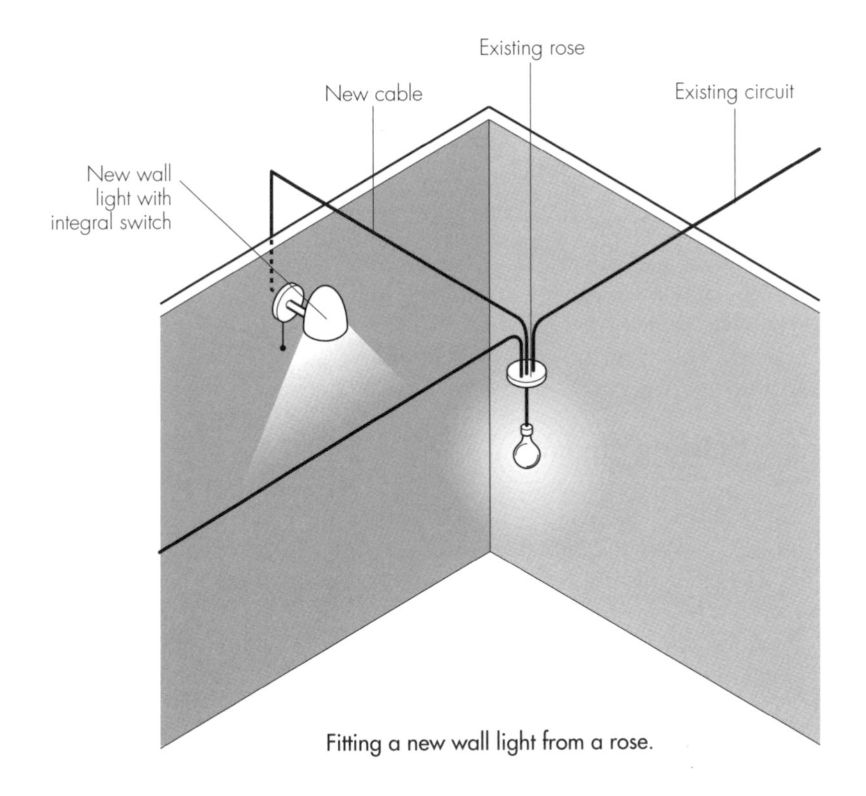

Fitting a new wall light from a rose.

extra cost of employing a plasterer if you want a seamless finish to the job. As soon as the plaster is dry, you can fix the light fitting to the wall.

The light fitting will come with instructions on how to fix it to the wall. Follow these carefully to make sure the fitting is properly secured. Most wall lights have a 3-pin terminal connection block and are pre-wired so that all you have to do is connect the new cable in the wall to the terminal block. Strip back about 50 mm of outer insulation and trim the excess. Strip 10 mm from the ends of each core and slide a length of earth sleeving over the bare earth core to leave about 10 mm protruding. See page 54 for the section on "Stripping cable". Using the electrician's screwdriver, connect the cable into the terminal block - live to live, neutral to neutral and earth to earth. If the lamp is made of metal, do a continuity test to check that the metal body of the lamp is connected to the earth terminal pin by using a multimeter - this should have been done in the factory, but you never know! If by any chance the fitting needs an earth strap, see "Earthing" on page 76.

Extending a lighting circuit

Before you start

Old houses, new houses... there never seem to be enough sockets and lights! And of course, a single light hanging in the centre of a room is sometimes just not enough, especially in a large lounge or a long kitchen.

Although extending the lighting circuit is quite a straightforward task, the disruption involved in lifting floorboards and redecorating is something else! You need to plan the operation carefully before you start work, deciding where to put the furniture and how to manage all the tasks or normal daily life while the lighting is turned off and through the general disruption.

You shouldn't have to go far to tap into an existing lighting circuit because every room will already be on one. Look for the nearest ceiling rose or the lighting junction box in the space above the ceiling.

Tools and materials

You will need the following tools: screwdriver, pliers, side cutters, wire strippers, earth sleeving, red/brown insulation tape, 4 mm electrician's screwdriver, electric drill, drill bits.

You will also need sufficient 1 mm^2 twin-core and earth cable to extend from the nearest ceiling rose or lighting junction box to the new location, to the new light fitting and to the wall switch.

Method

Switch off the circuit at the MCB (remove the fuse) and check that the circuit is DEAD. Locate the existing lighting circuit, which may be in the loft or under the floorboards above the ceiling. Lighting cable is usually the smallest size in the house, generally being only 1 or 1.5 mm^2, and the size is stamped into the outer insulation of the cable, so it should be easy enough to find. It is best to track the cable back to find both ends so that you know exactly

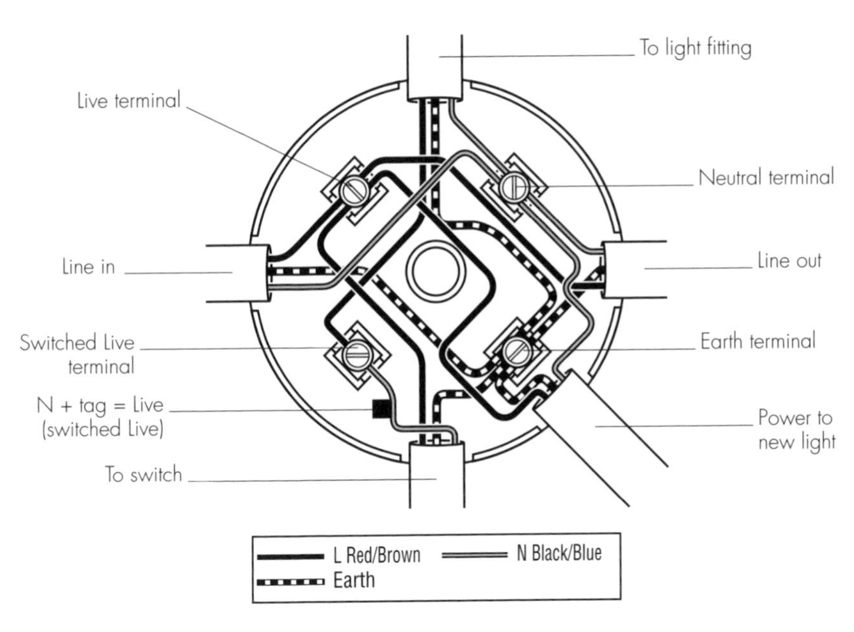

An extra light off a lighting junction box.

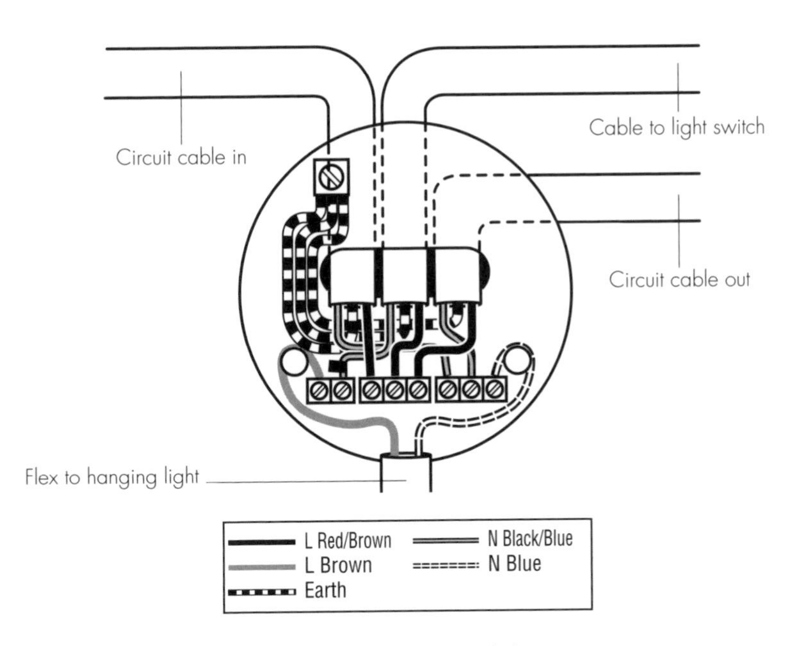

How a loop-in rose is cut into a lighting circuit.

where it is going. The appropriate cable will be connecting two ceiling roses or two lighting junction boxes.

Now you will need to add a junction box connection on the most convenient cable run. Check that the main power supply is turned OFF and cut the lighting cable. Strip about 50 mm off the outer insulation on the cable and strip about 15 mm of insulation off the cores. Slide a length of green+yellow earth sleeving over the bare earth cores to leave about 15 mm protruding. Refer to the section on "Stripping cable" on page 54.

From the new junction box run the twin-core and earth cable to the switch position you have selected on the wall or ceiling. If you are fitting a wall light, run another length of cable to the correct position on the wall (see page 58). If you are fitting a hanging pendant light, run the second cable to the exact position for the ceiling rose on the ceiling.

The procedure for wiring up a lighting junction box is much the same as for wiring up a loop-in ceiling rose. You will have four cables, two from the original circuit and the new ones that go to the switch and the light fitting. First wrap the black/blue wire that goes to the switch with red/brown insulation tape because this will be live when the switch is turned on. Next wire up all the cores to the appropriate terminals - all the black/blue neutral cores go to one terminal and all the green+yellow earth cores to another. The red/brown live core to the light and the black/blue live with the red/brown tag that goes to the switch must be connected to a single terminal by themselves. The remaining three red/brown live cores go to their own separate terminal.

If the cable to the light is to be connected to the flex from the light fitting in a ceiling rose, you need simply to connect live to live, neutral to neutral and earth to earth, unless the light fitting is made entirely of plastic, in which case there will not be an earth wire in the flex. To read more detail about making this type of connection, see page 74.

Changing single to two-way switches

Before you start

It is usual to fit a two-way switching system in hallways and stairwells so that the light can be turned off from one of two switches at each end of the space. Similarly, rooms that have two doorways often have a two-way lighting system. The wiring from a two-way switch to the ceiling rose and to the rest of the lighting circuit is the same as for a single switch; only the fact that there is wiring between the two switches is different. It is therefore quite easy to turn a single switch into a two-way switching system.

Tools and materials

You will need the following: a screwdriver, some red/brown insulation tape, pliers, side cutters, wire strippers, some earth sleeving, a 4 mm electrician's screwdriver, a short piece of earth core. In addition, you will need 1 mm^2 3-core and earth cable, which has a blue, yellow, red/brown and a

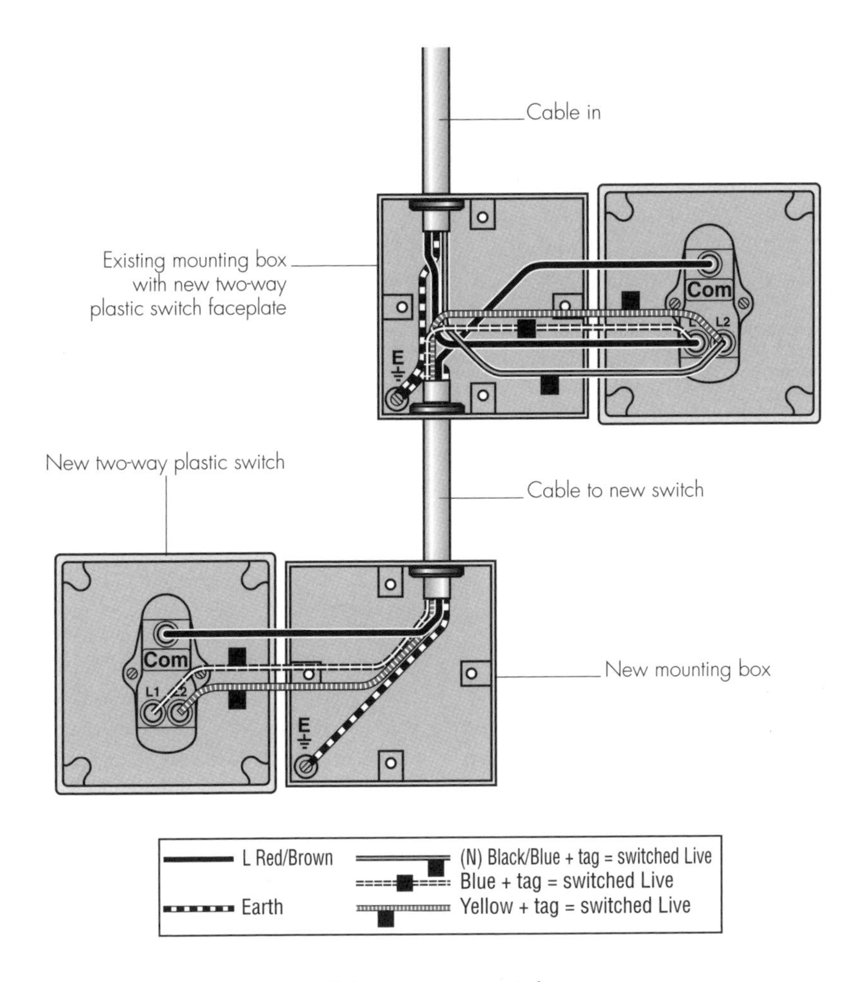

Fitting a two-way switch.

bare earth core in it, and probably general decorating equipment to make good the damage caused when fitting the new backing boxes.

Method

Switch off the circuit at the MCB (remove the fuse) and check that the circuit is DEAD. Open up the existing light switch. Inside you will see a single cable with red/brown live, black/blue (also live) and green+yellow earth cores connected to the terminals. The switch plate itself will probably have two connections, labelled "COM" and "L1", in addition to the normal earth terminal. If the black/blue wire does not have a red/brown insulation tape tag or sleeving to show that it is live when the switch is used, wrap tape around it now. Be aware that if the black/blue core is not tagged in the room, it might not be tagged in the loft and you should remedy this situation because a live black/blue switch core can be just as dangerous as the red/brown. Detach the cores and slip off the switch plate.

Refer to the section on "Running cable under floorboards" on page 62, and install ducting or surface-mounted trunking from the existing light switch to the position of the new two-way switch, taking it along the shortest vertical or horizontal route. Use a masonry drill, hammer and bolster to chop out a hole for the backing box. Fix the box with screws and wall plugs. In a plasterboard wall, clip a flush-fit box into a hole cut with a jigsaw.

Run the 3-core and earth cable through the new ducting from one switch to the other. Strip about 100 mm of outer insulation off both ends of the cable and trim the excess. Strip 10 mm of the insulation off both ends of all the single cores. Slip a length of green+yellow sleeving on the bare earth core as usual. If in any doubt about stripping the cable, see page 54.

Take the cable that comes from the ceiling rose to the existing switch position and wire it into the new two-way switch plate so that the red/brown live core goes to terminal L1 and the black/blue core with the red/brown tag goes to terminal L2.

Now take the cable that runs between the switches and connect it to the same switch plate. The red/brown live core goes into the COM terminal, the blue core goes into terminal L1 and the remaining yellow core goes into terminal L2. Tag both the blue and yellow wires with red/brown insulation tape since they can be live depending on the switch position. The earth cores all go to the backing box earth terminal, with a short earth strap going from the backing box to the switch plate if it is made of metal. Make the earth strap from earth wire and green+yellow sleeving.

Now wire up the newly-installed two-way switch. The red/brown live core goes to the COM terminal, the blue to L1 and yellow to L2. As before, both the blue and yellow wires should be tagged with red/brown insulation tape. The earth wire in this new switch goes to the backing box terminal as normal, with a short earth strap to the faceplate if it is metal.

Before you close up the switch plates, test the system. You will find that both switches turn the light on and off. Remember, however, that the usual saying that "up is off" does not apply to two-way switches. So… if you wish to change a light bulb, always turn off the main power supply at the consumer unit/fuse box.

Changing single sockets to doubles

Before you start

Changing a single to a double socket is a nice easy job, which can really tidy up your house. It gets rid of those nasty adapter plugs and strip plugs on extension leads! While it is not so long since most of us had only a radio and a TV, we now have all manner of electrical appliances, from extra TVs and video recorders to computers, printers, scanners, fax machines and so on - all needing sockets!

If your sockets are of the modern square 3-pin type, then they can easily be changed to double sockets. The wiring work is easy and does not require making any major changes to the circuit. Unfortunately, if you have the old-fashioned round pin sockets that are only suitable for 5-amp loads, you may not be able to convert them and will have to have the entire power circuit replaced.

Tools and materials

You will need the following tools and materials: new double socket fronts with surface-mounted or flush-fitted backing boxes, screwdriver, 4 mm electrician's screwdriver, electric or battery-operated drill, hammer, chisel, wall plugs, screws, pliers, 100 mm earth core and sleeving, as well as decorating equipment to make good the damage to the wall.

You will need to use a drill to complete this job, so you either need to run an extension lead from a neighbour's house or use a battery-operated drill because, of course, you won't be able to turn on the power until the job is done.

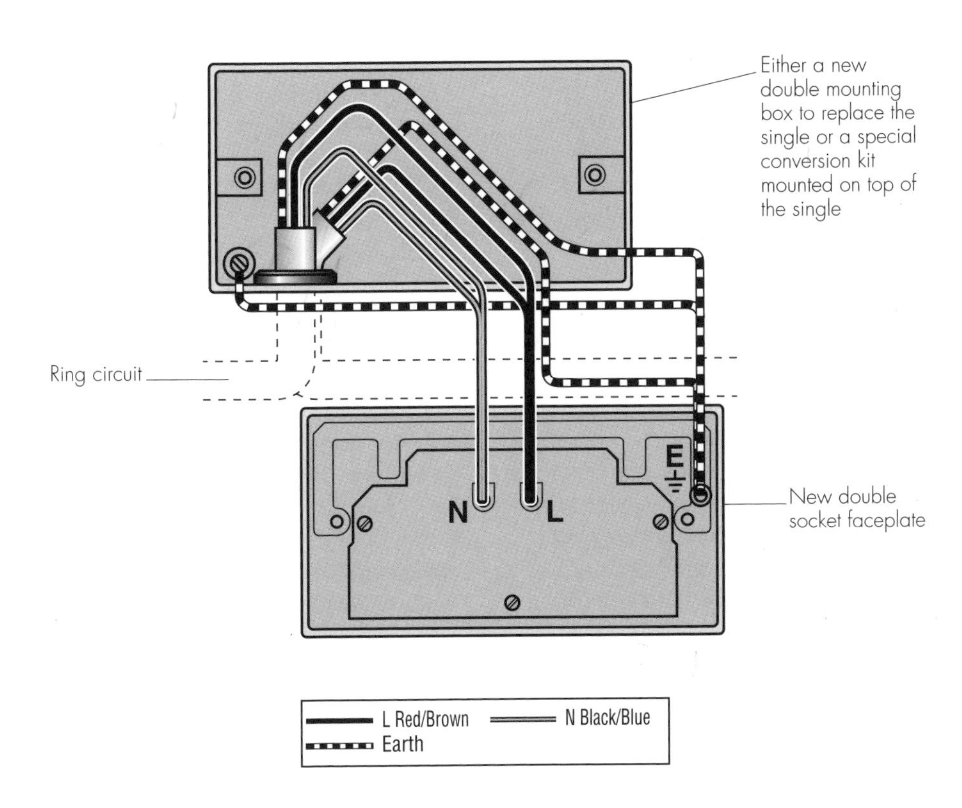

Changing a single socket to a double.

If you want to fit a double socket into plasterboard, specially designed easy-fit flush backing boxes are available at electrical suppliers.

Method

The easiest way to convert a single to a double socket is by fitting a surface-mounted backing box because the new box will either simply replace a surface-mounted single box or can be fitted directly over a single flush-fitted box. If you want the new double backing box to sit flush to a solid wall, first you will need to enlarge the hole in your wall.

Switch off the circuit at the MCB (remove the fuse) and check that the circuit is DEAD. Remove the existing faceplate and disconnect the wires from the terminals. If you are fitting a surface-mounted backing box, secure it using wall plugs and screws. To enlarge the hole to take a double backing box in a solid wall, first measure up the space needed for the backing box and drill away as much material as possible. Then use a hammer and chisel to clean out the hole so that it will take the box. If you're fitting a flush-fitted double socket specially designed for a plasterboard wall, cut the larger hole and then simply click the box into place. While you are working, and especially when chiselling and drilling, take care not to damage the exposed cables because you don't want to have to replace the cabling in the walls.

Once the backing box is securely in place you can fit the new double socket faceplate. You might need to use pliers to straighten the copper ends. Put all the red/brown cores under the live terminal, all the black/blue cores under the neutral terminal and all the green+yellow cores under the earth terminal. You will also need to run a short flying lead, with the earth core and sleeving, from the earth terminal to the metal backing box.

Check that all the connections are tight and cannot be pulled out from under the terminals. There shouldn't be any bare copper showing around the screw terminals. Carefully push the socket faceplate into position so that the wires sit tidily in the backing box without covering up the two fixing lugs. Using the two screws supplied with the socket faceplate, fix it securely without over tightening the screws, which might crack it.

Adding extra sockets

Before you start

The easiest way to create a few extra sockets is to replace single sockets with double (see page 90). However, sockets in new homes are usually doubles and this solution will not give you a socket in a new location.

If you want to add a totally new socket, first plan the work. You need to find a suitable existing single or double socket on the ring main to which you can connect a spur. The spur is the single cable going to your new socket. A ring main starts at the fuse box and passes in and out of a number of sockets before returning to the fuse box. This is because the sockets are fed power from both the in and out cables, which means there can be less wiring and more sockets on the circuit.

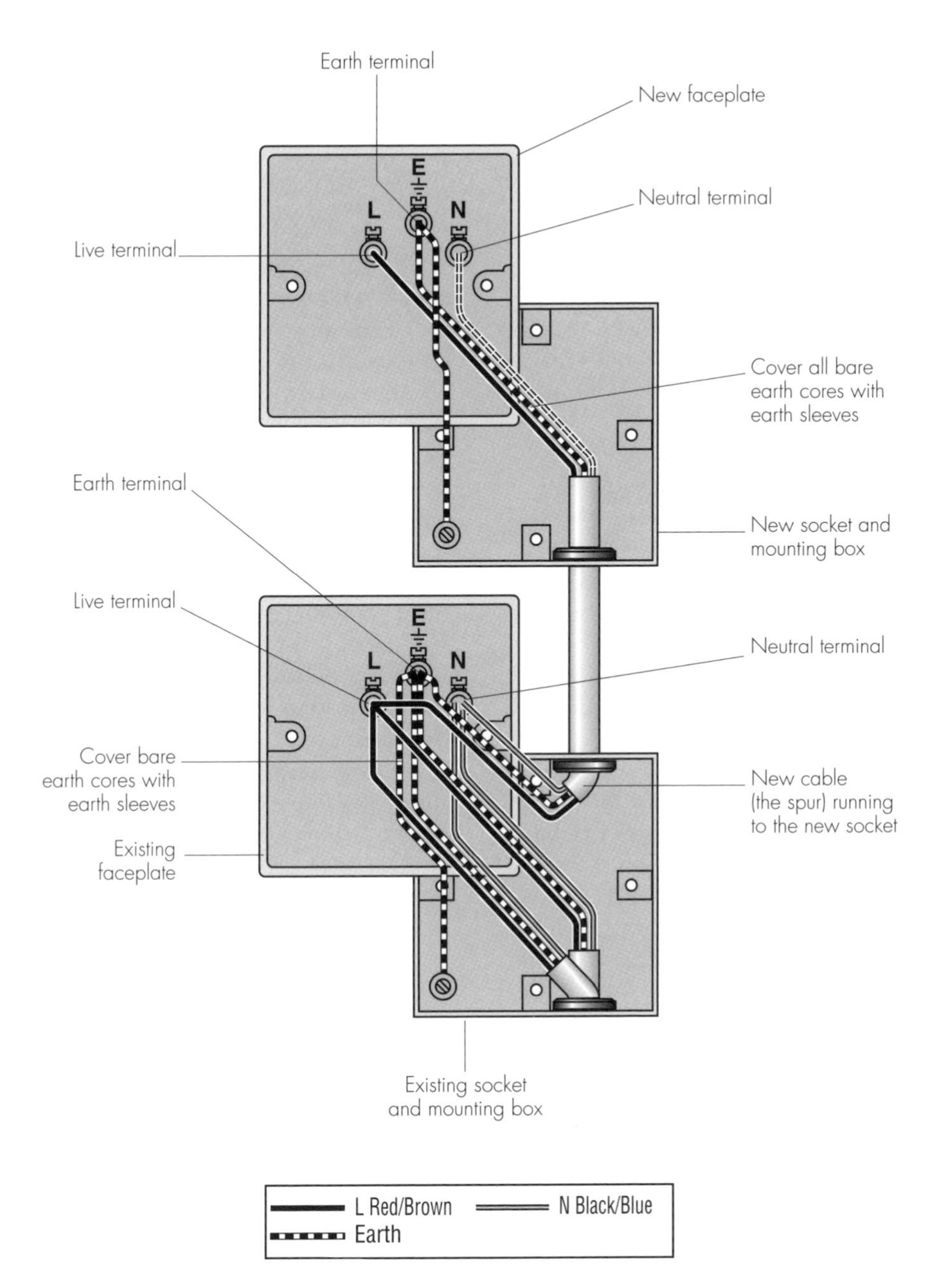

Wiring a new socket via a spur from an existing socket.

Tools and materials

When adding extra sockets you will need a length of 2.5 mm^2 twin and earth cable, long enough to go from the appropriate place on the ring main to the position of the new socket. This cable is flat, grey and has solid copper cores. You will also need a new socket and back box, a short length of earth core, some earth sleeve, a screwdriver, pliers, a masonry drill, wall plugs (No.6), screws (No.6 x 30 mm) and a continuity tester. A continuity tester is usually incorporated into a multimeter and is the safest tool available for testing electrical circuits. The tester should cost no more than a new CD player and should measure at least alternating current (AC), voltage (up to 1000V) and resistance, and have a continuity buzzer. The best place to buy a multimeter is a DIY superstore.

Method

Switch off the circuit at the MCB (remove the fuse) and check that the circuit is DEAD. When you are sure that the power is off, unscrew the socket from which you wish to extend the spur and make a small drawing of the wiring and connections with felt-tip pens. This will enable you to rewire it properly if you hit a snag.

Now consider if the socket is suitable for the job. If there are already three flat cables entering the box, you cannot use it because it already has a spur and a new connection would overload the wiring. If there are only two flat cables entering the box, it can be used for adding your spur. If there is only one flat cable, it could be either an existing spur, which you cannot use, or the last socket on a radial circuit, which is suitable for connection. To make sure which of these it is, track the cable back to the next socket with the continuity tester. If the cable goes back to a socket with two flat cables, it is on a radial circuit and it's fine to add your new spur. If the cable goes back to a socket with three flat cables in the box, it is itself a spur on the ring main and you cannot add another socket to this. This seems complex, but it is a simple process of elimination - so follow the rules above carefully and you will be fine. Once you have found an appropriate socket to connect into, close it up and turn the power back on. You can then mount the new back box on or in the wall and cut in the cable run.

You will need power for the drill for this, so do all of the work you can before turning off the main power.

Strip back about 100 mm of the new cable. The best way to do this is to grip the very end of the earth core (the bare, centre one) with pliers and pull it down the cable, opening the grey covering like a zipper. The 100 mm of grey covering can then be cut off. Next, strip 20 mm off both the red/brown and black/blue insulation. Then, using pliers, fold each end of the copper wire in half to make 10 mm ends. This must always be done when there is only one copper wire under a screw terminal to prevent the wire from being easily pulled out. Cut an 80 mm length of the earth sleeve, slip it over the earth core and fold the end of that wire too. Now cut a short length of earth cable and prepare either end as before. This earth tie is used to connect the socket earth to the box earth, as shown in the diagram (see page 91). Connect all the wires to the correct terminals, check everything and close up the new socket.

Now switch off the circuit at the MCB (remove the fuse) and check that the circuit is DEAD and feed the other end of the spur cable into the existing socket box. Prepare the cable as before, but with 10 mm bare ends. Do not fold the ends over, since this cable shares terminal screws with the other incoming cables. Wire up all the reds/browns to L, all blacks/blues to N and all earths to E. Make sure there is a 100 mm earth tie from the socket front to the box – it should already be there if the socket was properly wired originally! Give all the wires a good pull to check that none are loose and close up the socket.

When you turn the power back on at the main switch, no breakers should trip or fuses blow. To test the system, plug lamps into the new socket and the one you connected into. If one of the sockets does not work switch off the circuit at the MCB (remove the fuse) and check that the circuit is DEAD, open the sockets you have been working in and check the terminal screws are tight. All the bare wire ends should be right into the terminals up to the insulation.

Fitting fused connection units

Before you start

Fused connection units (known as FCUs) are very useful for permanently connecting appliances to the main power supply. They cannot easily be disconnected, unlike ordinary sockets, so are ideal for domestic appliances such as fridges, freezers and washing machines.

An FCU is rather like a single socket and is installed in a similar way since it fits on a standard flush-fitted or surface-mounted backing box. There are two types of FCU. One has a cable-in and cable-out connection, whereas the other type has a cable-in and flex-out connection. The cable-in and cable-out design is used for extending a power or lighting circuit, whereas the cable-in and flex-out unit is used for connecting appliances that have a flex. Both types are available as switched or unswitched units, so choose whichever one is best suited to your needs.

The easiest way to connect an FCU into the electrical system is to find the nearest socket and run a short spur of 2.5 mm^2 cable to the FCU. You cannot use a cable lighter than 2.5 mm^2, as otherwise it will not be protected by the MCB or fuse in the consumer unit or fuse box. When you are deciding on the best place to locate the FCU, try to find a position that is out of sight, such as behind the washing machine, dishwasher, refrigerator or other appliance, so that the flex and FCU are safely hidden – well away from children.

Tools and materials

You will need the following: screwdriver, tools and materials for running the cable and fitting the backing box, a length of 2.5 mm^2 cable, string or wire mouse, pliers, side cutters, wire strippers, 4 mm electrician's screwdriver, earth sleeving, the FCU unit of your choice, the appropriate fuse.

You will need the correct fuse to put in the little socket on the front of the FCU. You need the same type of fuse as is found in 13-amp plugs. Use a 3-amp fuse for an appliance needing up to 700 W, a 5-amp fuse for 700 to 1200 W, and a 13-amp fuse for more powerful appliances needing over 1200 W, such as a washing machine. Once you have cut the plug off the flex from the appliance, you could fit the fuse from the plug in the FCU fuse holder.

Method

Switch off the circuit at the MCB (remove the fuse) and check that the circuit is DEAD. You will need to run ducting or trunking from the nearest socket to the site for the new FCU. Before that, take off the socket faceplate so as to avoid damaging the socket and wiring. If you have sufficient space behind the appliance for a surface-mounted backing box, use this type of FCU as it will be much easier to fit and cause less damage to the wall. For mounting cable in trunking, see page 60. However, many appliances take up the exact depth underneath a worktop, so you may have no option but to fit a flush-fitted backing box. For running cable in ducting, see pages 64–65.

Run the cable through the ducting using string or wire as a mouse. Strip about 100 mm of insulation off both ends of the cable, referring to page 54. Then strip about 10 mm off the red/brown and black/blue cores. Cut about 80 mm of earth sleeving and slide this over the bare earth core.

On the back of the FCU front plate you will see two sets of connections, one set labelled "LOAD" and the other set "SUPPLY". Connect the cable from the socket to the SUPPLY terminals, black/blue core to neutral, red/brown core to live and green+yellow core to earth.

Connect the other end of the cable to the original socket. On the back of the socket you will see all the red/brown wires under the live terminal, and all the black/blue wires under the neutral terminal. Loosen the terminal screws on the back of the socket faceplate and connect the cores from your spur cable – red/brown live with the other reds/browns under the live terminal, the black/blue with the other blacks/blues under the neutral terminal and the green+yellow core with the other earths under the earth terminal. Make sure all these connections are secure before replacing the faceplate and fitting the cover screws.

If your FCU is the type with a flex output, you will see a cord grip on the back. Loosen the screws and remove the cord grip. Cut the plug off the flex from the appliance. Strip about 100 mm off the outer insulation on the flex, taking care not to damage the coloured insulation on the cores. Then strip about 20 mm of insulation from each core. Twist the fine copper

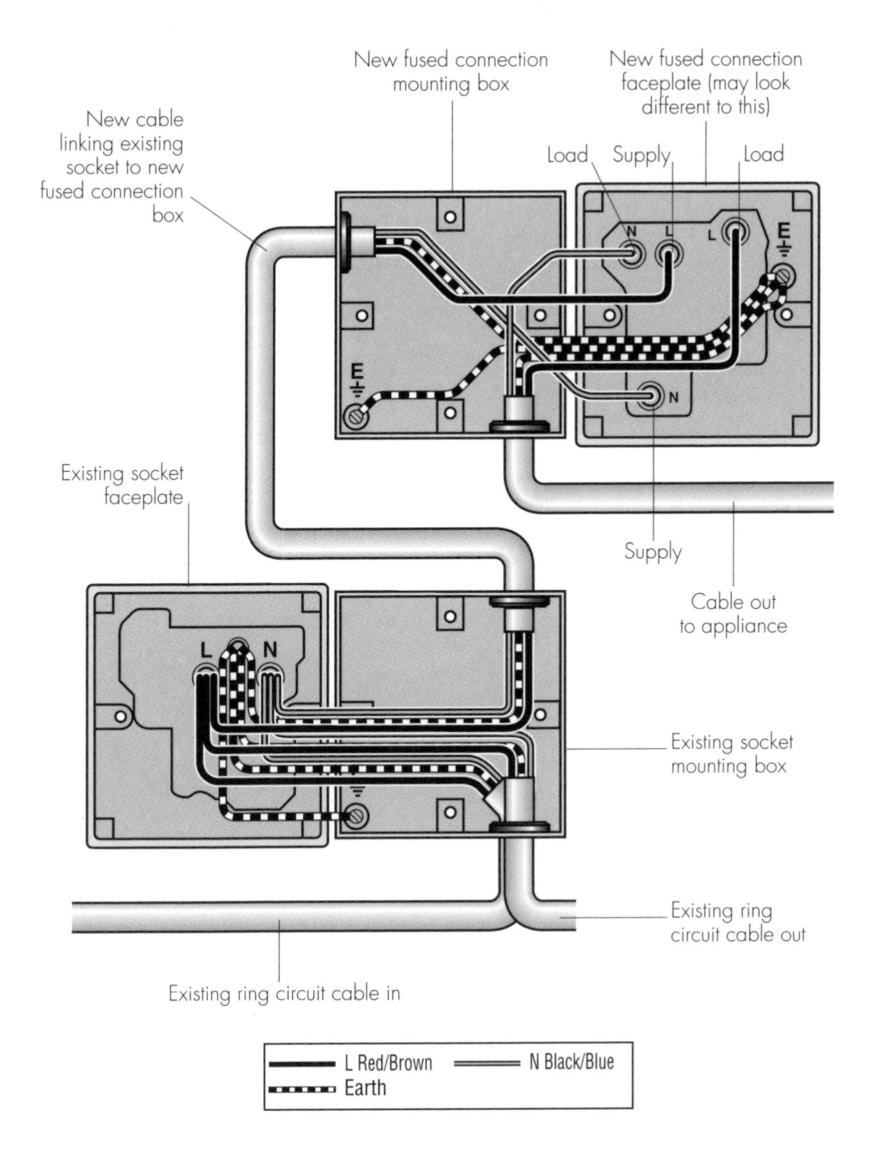

Fitting a fused connection unit.

wires with your fingers so that they don't fray and fold each bare copper end in half. For detailed instructions refer to "Stripping cables" on page 54.

WARNING: don't leave the old plug with a short length of flex attached lying around. It's a potential killer as it's just the sort of thing children might try to push into a socket! Once you have cut the plug off the appliance, immediately either remove the flex and keep the plug or break the plug with a hammer and throw the whole thing away.

Connect the flex under the LOAD terminal screws - brown to live, blue to neutral and green to earth. Check that all the connections are sound by giving them a pull and then replace the cord grip and screws. Make sure that the flex cannot be pulled out from under the grip.

If the FCU has a cable outlet, strip the layers of insulation in a similar manner to the flex and connect the cores under the LOAD earth, neutral and live terminals. Make sure these connections are sound and then screw the cover back into position.

To complete the job, fit the correct fuse in the FCU fuse holder.

Wiring up an extractor fan or cooker hood

Before you start

It's good that houses now usually have double glazed windows and doors since it stops draughts whistling around, but the downside is that bathrooms and kitchens are very prone to condensation as a result. The cure is to fit an extractor fan in the bathroom and an extractor fan and/or a cooker hood in the kitchen.

Extractor fans can easily be fitted on bathroom walls or ceiling spaces and are already built into many cooker hoods. If a fan cannot be positioned on an outside wall, you will need to run a 100 mm diameter concertina vent pipe from the fan to the outside. If the fan is a ceiling-mounted unit, the vent pipe will need to be run through the loft and out under the edge of the roof. If there is a fan already fitted on an outside wall, you will be able to run a short piece of 100 mm diameter plastic pipe straight through the wall to the grill on the outside. Extractor kits, including the vent pipe, can be bought in most DIY stores. There are also all manner of designs of grill that can be put on the outside of the wall to cover the pipe.

You also need to decide how you want the fan to operate. You can choose to have it operate from the existing light switch or pull-cord, or you can have a separate switch or a pull-cord to turn the fan on independently of the lighting. Remember, you must use a pull-cord in a bathroom, toilet or any other wet room. Usually a fan in a wet room is wired directly into the lighting, so that the fan comes on automatically when the light-cord is pulled on.

However, if you want a separate switch you must fit a double-pole pull-cord ceiling switch between the fan and either the lighting circuit or an FCU, if connecting into the socket circuit. Some fans already have a pull cord but this can often end up in an inconvenient position so using a double-pole switch also has the advantage of making the switch more accessible.

Cooker hoods that have a fan and a very simple charcoal filter have a push-fit connection at the back to take the 100 mm diameter vent pipe.

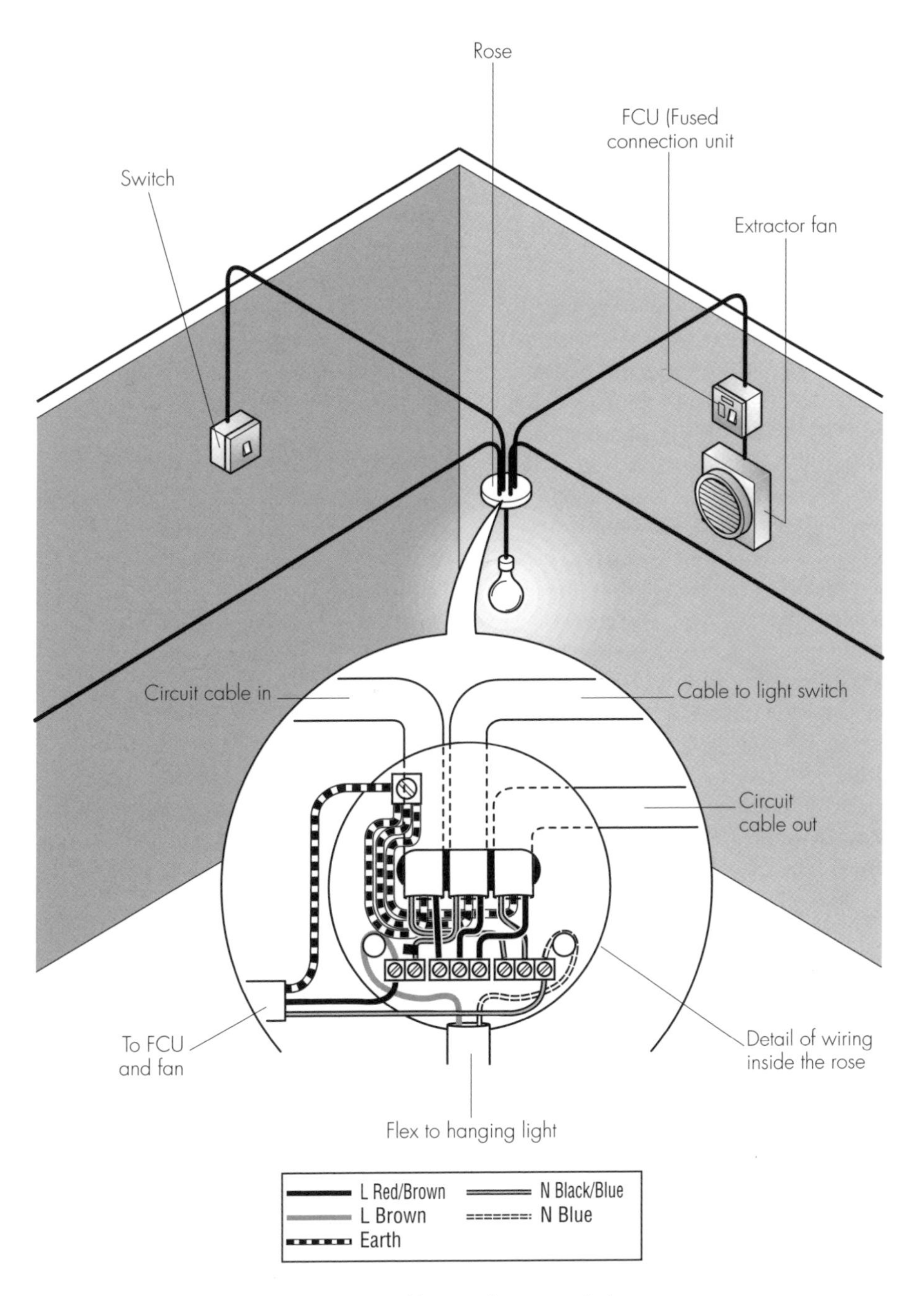

Fan activated by switching on a light.

Connecting a fan to a lighting circuit

You will need the following: flat-head screwdriver, pliers, side cutters, wire strippers, 4 mm electrician's screwdriver, 1 mm^2 twin-core and earth cable and, if the fan does not have a switch of its own, an FCU with a backing box.

Fix the fan to the wall or ceiling using the manufacturer's instructions. If you need an FCU to operate the fan, position it where you can easily reach the FCU. Switch off the circuit at the MCB (remove the fuse) and check that the circuit is DEAD. Connect the FCU, referring to page 95, and connect the fan into the FCU. Run a length of cable from the FCU, or directly from the switched fan, to the ceiling rose or lighting junction box, depending on how your lights are connected. Connect the cable from the FCU under the same terminals as the cable from the light fitting – so that the fan comes on when you switch on the lights. See the diagram on page 87 for connection to a junction-box system and for connection to a loop-in rose.

Connecting a switched fan to a lighting circuit cable

You will need the following: flat-head screwdriver, pliers, side cutters, wire strippers, 4 mm electrician's screwdriver, 1 mm^2 twin-core and earth cable and a 3-terminal round junction box.

Switch off the circuit at the MCB (remove the fuse) and check that the circuit is DEAD. Find the lighting circuit cable in the loft or ceiling space. It will be a 1.00 or 1.5 mm^2 twin-core and earth cable with the size stamped on the insulation. Check that it supplies power to the nearest ceiling rose or light fitting. Cut the cable and prepare the ends by stripping about 60 mm off the outer insulation. Fix the junction box to a beam or joist and connect both the lighting cable and the cable from the fan. Connect all the red/brown lives under one terminal, all the neutral blacks/blues under another and all the green+yellow earths under a third.

Connecting a fan into a socket circuit

You will need to run a 2.5 mm^2 spur cable from the nearest socket to the position you have selected for the FCU, referring to page 95. Connect the fan into the FCU in accordance with the manufacturer's instructions.

Fitting dimmer switches

Before you start

It is a nice option to have a dimmer switch in a living room or bedroom, so that you can control the level of lighting from bright light to a subtle glow to suit your moods and needs. Dimmer switches have improved since the days when they produced lots of buzzing noises and sometimes overheated. With the introduction of new heavy-duty transistor technology, dimmer switches are now long-lasting, safe and reliable.

Changing an ordinary switch to a dimmer switch is quite easy and is the DIY task that anyone should be able to do. Unfortunately, if your room has a two-way switching system, where more than one light switch controls a light fitting, then you will not be able to use a dimmer. In addition, dimmer switches only work with resistive loads, meaning bulbs with the traditional filament wires, and therefore cannot be used with fluorescent lights, either the long tube strip lights or the new energy-efficient bulbs. To check if a light bulb is of the fluorescent energy-efficient type, take it out and inspect it. Fluorescent bulbs have a large white plastic base, whereas filament bulbs tend to have small metal bases.

You can choose dimmer switches with the usual white plastic faceplates or from a wide selection of decorative finishes, such antique brass or stainless steel. You can also buy dimmer switches that come with a small remote control unit so you can turn the lights up or down from the comfort of your armchair. These remote control dimmers are quite expensive but excellent if you have trouble getting out of a chair or just want to show off!

Tools and materials

You will need a flat-head screwdriver, wire strippers, a 4 mm electrician's screwdriver, 100 mm of earth cable, 80 mm earth sleeving, the dimmer switch of your choice.

Method

Switch off the circuit at the MCB (remove the fuse) and check that the circuit is DEAD. Unscrew the cover screws on the faceplate and remove it. Inside the switch you will find three cores - the green+yellow earth that is connected to the backing box, the red/brown live core and the red/brown tagged black/blue live core. Loosen the terminal screws and detach the cores. If the cores are in good condition, connect them to the new switch. If the copper ends look as if they might break off, strip them back and then make the connection.

If the faceplate is metal you will need to fit a short earth strap linking the faceplate to the backing box. Slip the sleeving over the short length of earth cable and make the connection. If there is no earth terminal in the backing box, a plastic-fronted dimmer switch will be better than a metal one.

Ease the wires carefully into the backing box and screw the faceplate into position. The new dimmer switch may have come with its own screws, but it's just possible that these will not screw into the backing box and you will have to use the original screws.

Fitting central heating controls

Before you start

Central heating controls comprise a clock and switches that are designed to turn heating on and off at specific times and to regulate the temperature in the rooms. While in theory they are fairly basic, there are many different designs using quite idiosyncratic wiring colours and terminal designs. Probably the best advice is to leave this work to a qualified electrician.

However, if you study the system, make a diagram and have a good knowledge of electrical circuit theory, there is no reason why you cannot replace central heating control components yourself. Take great care when changing any unit and always follow the manufacturer's instructions very carefully to the letter.

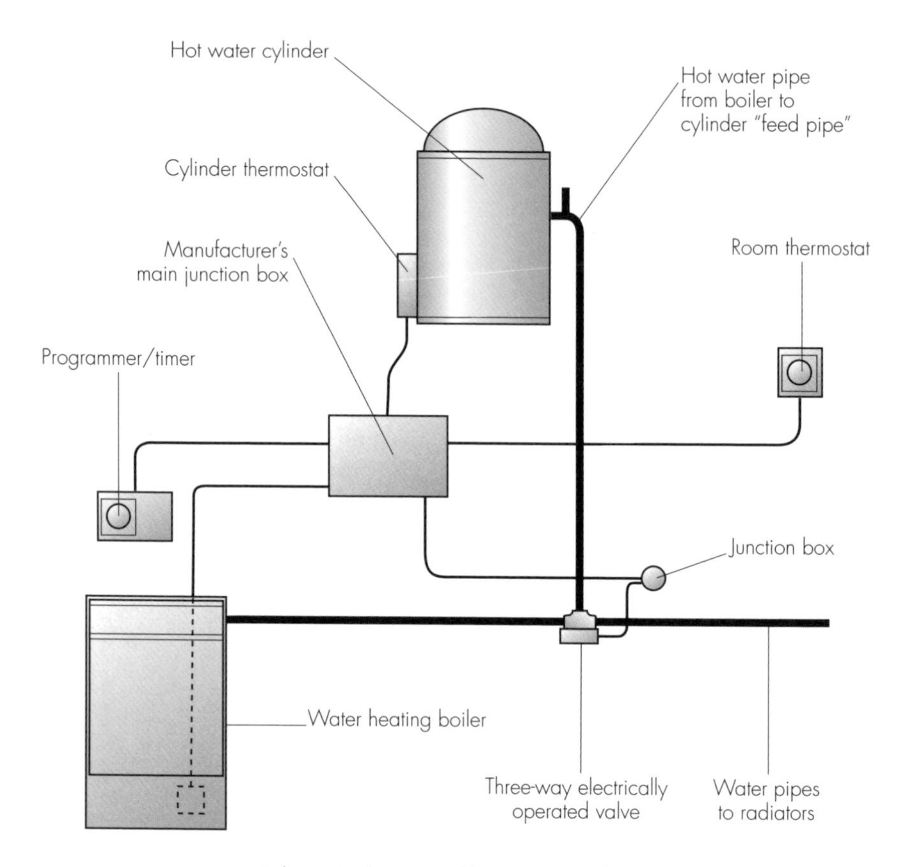

Schematic diagram of heating controls.

There are three main types of component employed in a central heating control system – the timer or programmer, the room thermostat, and a motorized valve and water cylinder thermostat. Most central heating controls require 230 volts usually supplied via an FCU connected to a socket ring circuit. If the central heating system has let you down in some way, the FCU is the first place to look. Make sure that the fuse is okay and the switch is in the on position. From the FCU the power usually runs to a main junction box supplied with the central heating controls. This box contains all of the connections from the different components of the central heating control system. You will need the manufacturer's circuit diagram (sometimes printed inside the lid) in order to find out the purpose of each terminal. The wiring in these types of connection box is quite individual, so you will need to follow the manufacturer's instructions.

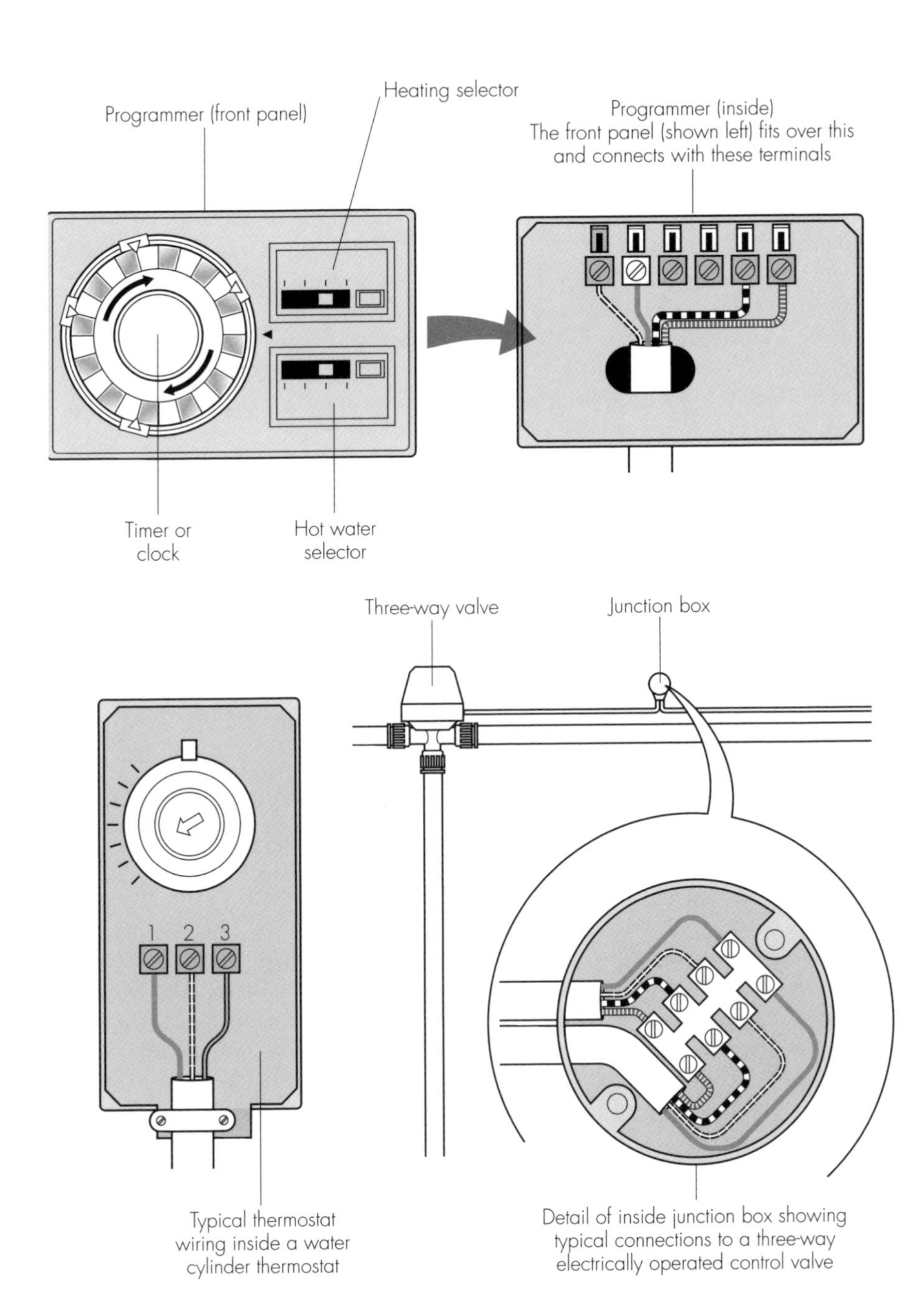

Schematic diagram showing the inside of heating controls.

Fitting components

This section gives general advice about fitting replacement programmers, room thermostats, cylinder thermostats and electric valves. For detailed instructions on installing the power supply, refer to "Fitting fused connection units" on page 95.

Fit the central heating programmer somewhere reasonably accessible, although out of the way of small children. It needs to be in a position with good lighting so that you can read the display or the clock numbers. Before removing the old unit, make a diagram of the original wiring. Most programmers have a circuit diagram on a label, so you should be able to work out the purpose of each terminal and find the equivalent in the new unit. Switch off the circuit at the MCB (remove the fuse) and check that the circuit is DEAD before starting work.

Room thermostats have either two or three wires connected to simply open and close the switch according to the temperature of the room. Don't fit the room thermostat above a radiator or beside a doorway because it will be adversely affected by the flow of cold and hot air.

Cylinder thermostats also have two or three wires. They can be found clamped to the side of the hot water cylinder, usually at a point low down and under the insulated covering.

The motorized valves in your central heating system are used to control the way the water flows around the pipes. Although such valves are fairly foolproof, they can stick and fail with age. The task of replacing a valve is best left to a plumber, but the wiring is a very simple task involving no more than wiring in four cores. It is best to fit a small 4-terminal junction box between the motorized valve and the central heating wiring to make it easier to replace the valve, which usually comes with a short tail flex already connected. Always make a diagram before you start work and, if in doubt about the connections, call in an electrician rather than a plumber.

Wiring up wall heaters and towel rails

Before you start

You are not allowed to have normal plugs and sockets in a bathroom, because water is a perfect conductor of electricity and consequently really dangerous. If you can sit in the bath or stand in a shower and reach a socket or a switch, then you are in a life-threatening situation!

If you need to fit an electric heater or light you must make sure that no part of the unit can be touched with wet hands, that is with the exception of a special bathroom towel rail. Towel rails come in many different designs and may have quite specific wiring requirements. Always follow the manufacturer's instructions for mounting a rail to the wall or floor and wire it up with the utmost care.

For reasons of safety wall heaters in bathrooms are not allowed to have the normal plastic light-type switch. If the wall heater has its own pull-cord switch with a pop-up on/off indicator, you will need to connect it to the FCU and then into the socket ring circuit or

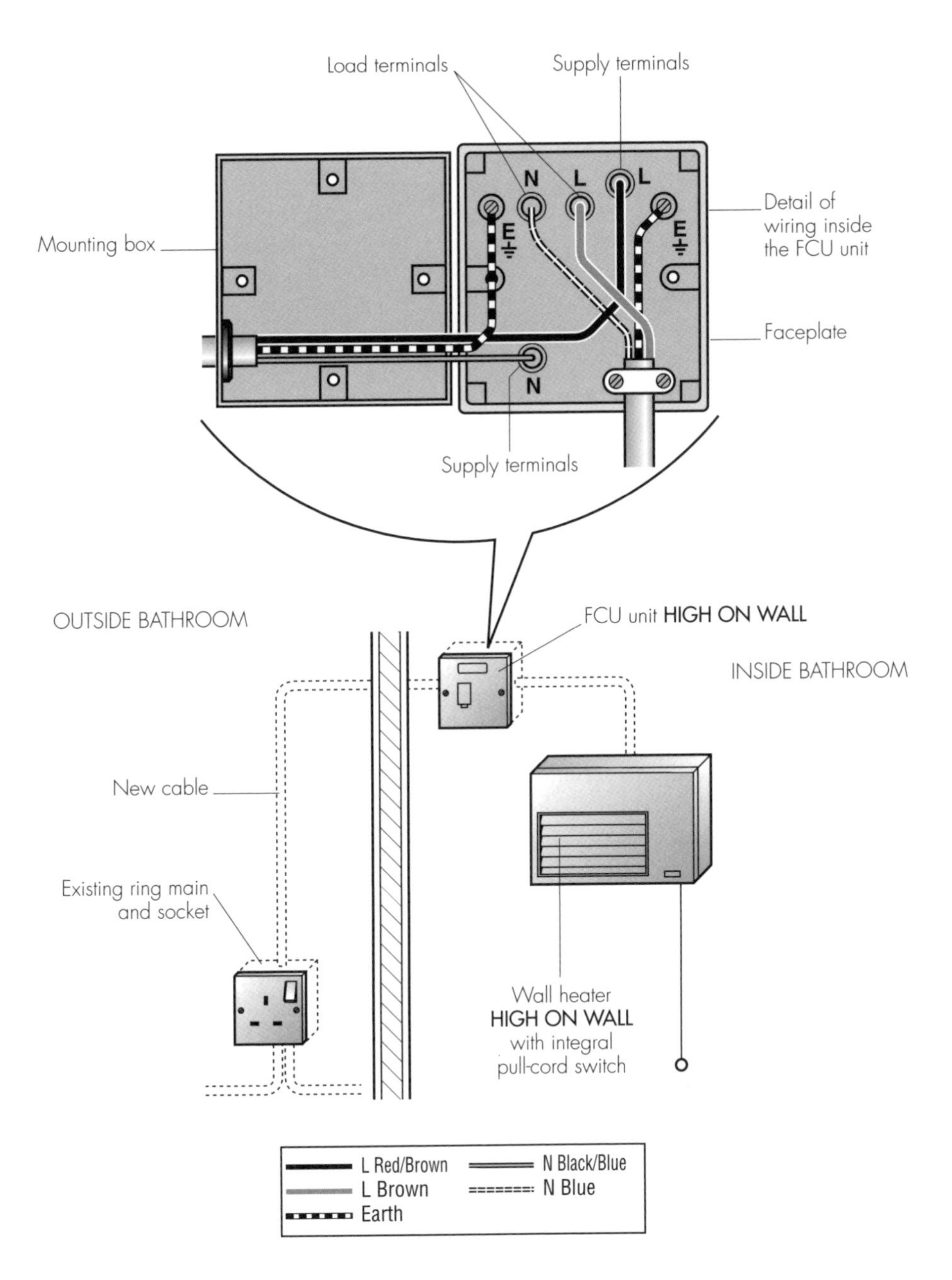

Bathroom wall heater via ring main socket.

to the fuse box. If the wall heater does not have its own switch, then you will need to fit a double-pole, ceiling-mounted, pull-cord switch between the appliance and the FCU.

Fitting a wall heater without an integral switch

You will need the following: flex outlet plate (3-terminal connector on a single socket sized box) with a flex hole, screwdriver, 2.5 mm² twin-core and earth cable, pliers, side cutters, wire strippers, 4 mm electrician's screwdriver, earth sleeving, double-pole pull-cord ceiling switch with a mechanical pop-up on/off indicator, FCU with just a fuse holder and red neon light to show it has power, not a switch.

First fix the heater to the wall in accordance with the manufacturer's instructions. Fit the flex outlet plate to one side of the heater. It will fit on a normal single backing box. Connect the shortest possible length of cable from the heater to the outlet plate. Then run cable from the flex outlet plate to the double-pole ceiling switch and from there to an FCU that is connected into the socket ring circuit. Refer to page 95 for connecting the FCU. It can be positioned inside the bathroom but it must be very close to the ceiling so that it cannot be touched.

In the flex outlet plate connect the blue and black/blue wires to the neutral terminal, red/brown and brown wires to the live terminal and the green+yellow earth wires to the earth terminal. The double-pole ceiling switch has five terminals. Connect the cable coming from the FCU to the SUPPLY live and neutral, and connect the cable going to the flex outlet plate to the LOAD live and neutral. In the FCU, connect the cable from the ceiling switch to the LOAD live and neutral terminals. Make sure that the earth wires are properly connected to the earth terminal. The pull-cord switch should be labelled 'Wall heater' so that it cannot be confused with the shower or light switch.

Fitting a wall heater with an integral switch

You will need the following: FCU with an integral red neon indicator, fuse holder and flex outlet, screwdriver, 2.5 mm² twin-core and earth cable, 3-core 2.5 mm² flex, pliers, side cutters, wire strippers, 4 mm electrician's screwdriver, earth sleeving.

Mount the heater on the wall in accordance with the manufacturer's instructions. It should be positioned high on the wall so that only the pull-cord can be reached. Follow the fire safety instructions to the letter, especially those that tell you how far from the ceiling, curtains and floor the heater needs to be located.

Fix the FCU alongside the heater, again high on the wall so that the unit cannot be touched without the use of a step-ladder. There will be a hole for the flex either in the front plate or at the bottom. Refer to page 95 and run the cable from the socket ring circuit to the FCU and the flex from the FCU to the heater.

Make sure there is an earth

connection throughout your new wiring right from the wall heater to the socket ring main. When you have completed the work have it tested by an electrician.

Wiring up an immersion heater

Before you start

Even if you have central heating, it's still a good idea to have an immersion heater as a back-up and most houses have one. An immersion heater acts like a giant electric kettle element in your hot water cylinder. All you will be able to see on the outside of your water cylinder is a plastic dome or box, with a large hexagonal nut and a flex connecting it to a switch, timer or flex outlet plate.

An immersion heater uses a lot of electricity so it must be connected directly to the consumer unit or fuse box with a separate 20-amp miniature circuit breaker (MCB) or fuse. Sometimes, in the interest of economy, an immersion heater is connected to the "night time in" electricity meter. So don't be too surprised if you find that your immersion heater doesn't work in the daytime! There are several different ways in which you can wire up immersion heaters, the most common ones of which are covered here.

Fitting an immersion heater with a ceiling pull-cord switch

You will need the following tools and materials: 20-amp MCB or fuse, 20-amp double-pole ceiling switch with a mechanical pop up on/off indicator, flex outlet plate (3-terminal connector on a single socket sized box) with a flex hole, screwdriver, 2.5 mm^2 twin-core and earth cable, 2.5 mm^2 heat-resistant flex, pliers, side cutters, wire strippers, 4 mm electrician's screwdriver, earth sleeving.

Switch off the circuit at the MCB (remove the fuse) and check that the circuit is DEAD. Run cable from the new MCB or fuse in the consumer unit or fuse box to the ceiling switch placed conveniently near the immersion heater. Do not put the switch in your airing cupboard because you will be liable to forget whether it is on or off when you can't see the indicator. Run another length of cable from the ceiling switch to a flex outlet plate and the heat resistant flex from the flex outlet plate to the immersion heater. Connect everything up and make sure that the earth connection is continuous right from the immersion heater to the consumer unit.

Fitting an immersion heater with a double-pole switch

A simple and cheap way of controlling the immersion heater is by installing a double-pole switch. You will need the following tools and materials: 20-amp MCB or fuse, 20-amp double-pole wall switch with a neon on/off indicator (you will need a dual switch unit if you have a two-element immersion heater), flex outlet plate, screwdriver, 2.5 mm^2 twin-core and earth cable, 2.5 mm^2 heat-resistant flex, pliers, side cutters, wire strippers, 4 mm electrician's screwdriver, earth sleeving.

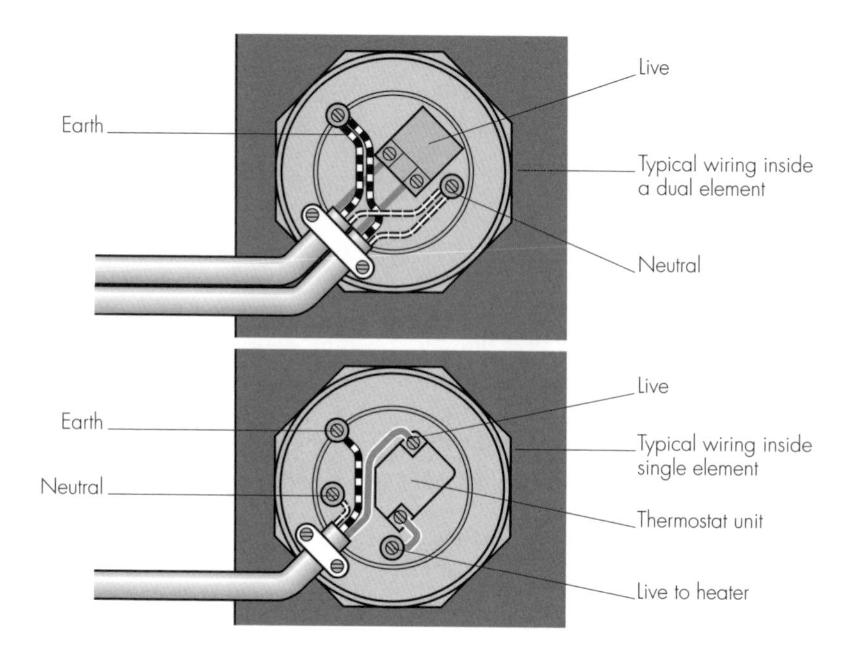

Immersion heater connections.

Switch off the circuit at the MCB (remove the fuse) and check that the circuit is DEAD. Run cable from the new MCB or fuse in the consumer unit or fuse box to the wall switch placed conveniently near the immersion heater. The ideal location for the switch is just inside in the airing or water cylinder cupboard well away from wet hands - certainly not in the bathroom. Make sure the earth connection is continuous from the immersion heater element to the consumer unit.

Fitting heaters with timers and clocks

Many immersion heaters have a small clock or timer to control the heating of the water. This is usually fitted between the double-pole switch and the immersion heater element. All clocks and timers require power for the clock mechanism to work, so if the double-pole switch is turned off the clock will stop. Make sure that the clock is set to the correct time, especially if you're using economy night-time electricity. The clock usually has five terminals, one of which will be clearly marked as earth. The other four terminals are for SUPPLY live and neutral, and LOAD live and neutral. Most clocks come with a wiring diagram, which may be on a label or moulded into the plastic, to show you how to connect the terminals. Clocks and timers vary in design, so be sure always to follow the manufacturer's installation instructions carefully.

Wiring up a cooker and hob

Before you start

Cookers and cooker/hob combinations are awkward to wire up because they require their own dedicated power circuit.

Before you start work you will need to calculate the size of cable and MCB or fuse needed for the cooker. Have a look on the back of the cooker and find the plate that shows the rating, or the number of watts the cooker uses. For example, the cooker might use 12 kW, which equals 12,000 W. If you divide the 12,000 watts by 230, you get a figure of 52.17 amps. This is not a particularly meaningful figure given that 50 amps is about the capacity of the entire electrical supply in a house and, because it is not likely that every part of the cooker will be on at the same time, it is conventional to make a calculation using the diversity principle. It is normally accepted that 10 amps are always required by a cooker with the addition of 30% of the remaining number of amps.

So, to continue the example, if you take 10 amps from the original 52.17 amps, you have 42.17 amps, 30% of which equals 12.65 amps. If you then add the 10 amps and the 12.65 amps together, you get a figure of 22.65 amps. Then add on another 5 amps for the 3-pin socket on the cooker switch to arrive at a grand total for the diversity load of 27.65 amps. Finally, you need to round this up to the next fuse size, which is 30 amps and the size of fuse you need for a 12 kW cooker. If you have a separate oven and hob, add the two ratings together and calculate the size of fuse using the same diversity calculation.

For a normal cooker, under 12 kW, use 6 mm^2 twin-core and earth cable if the circuit is up to 10 m long or 10 mm^2 cable if the circuit is more than 10 m long. For a bigger cooker, over 12 kW, use a 45-amp fuse and some 10 mm^2 twin-core and earth cable if the circuit is up to 6.5 m long, but 16 mm^2 cable for circuits of between 6.5 and 10 m long. If you have a separate oven and hob, you will be able to connect both appliances to the same switch as long as they are both within 2 m of the switch.

If you know that you will be using most of the capacity of a 12 kW cooker most of the time, for example if you are running a guesthouse or café, or if you cook in the home in some kind of professional capacity, you may wish to go up to a 10 mm^2 cable and 45-amp circuit breaker.

Tools and materials

You will need the following tools and materials: MCB or fuse of suitable size, some twin-core and earth cable of suitable size, a double-pole switch (the cooker switch) with/without an additional 13-amp socket and red neon indicator, a 3-terminal junction box, a screwdriver, pliers, side cutters, wire strippers, 4 mm electrician's screwdriver, some earth sleeving.

Method

Switch off the circuit at the MCB (remove the fuse) and check that the circuit is DEAD. Fit the new 32- or 45-amp MCB or fuse in the consumer unit or fuse box and run the appropriate size of twin-core and earth cable to the cooker switch. The switch must be positioned within 2 m of the cooker and hob, but must not be placed directly above them in case there is a flame licking up from the cooker. Connect the new cable to the SUPPLY live and neutral terminals in the cooker switch. Slip a length of green+yellow sleeving on the earth wire and connect it to the earth terminal.

Next run cable from the LOAD live and neutral terminals in the cooker switch to the three-way junction box for the cooker, which is usually immediately behind the appliance. Slip a short length of earth sleeving onto the earth core and connect the red/brown live, black/blue neutral and green+yellow cores to the appropriate terminals in the junction box.

If the cooker already has a cable connected, you can fit this into the junction box, connecting the cores to match those from the cooker switch. If there is no existing cable, then use the same size as you have for the rest of the circuit. In the back of the cooker you will find three connections for live, neutral and earth. The earth terminal is usually located under a small bolt or nut. Make sure the earth connection is continuous all the way from this bolt back to the consumer unit.

If you have a cooker and hob combination, you can run the cable from the junction box first to the cooker and then on to the hob, but the best way, as long as both units are within 2 m of the switch, is to connect both appliances into the cooker junction box. If the cooker is further than 2 m from the switch, you must fit two separate switches.

WARNING: NEVER fit flex on to a cooker – it will not carry the amount of current needed and will simply overheat and melt.

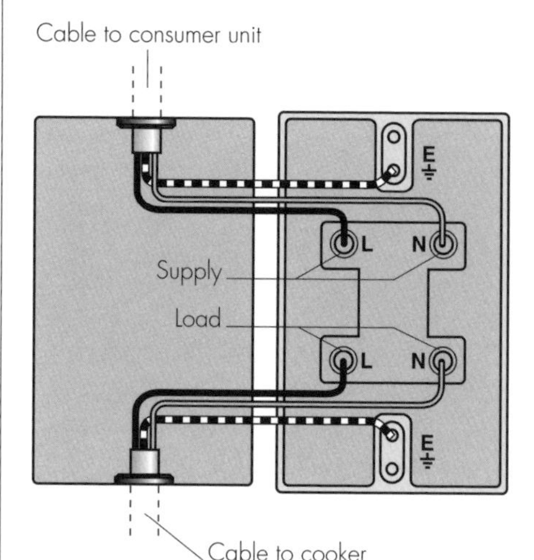

Connection at cooker control unit.

Wiring up an instant shower

Before you start

Instant electric showers are winners on at least two counts. They are relatively easy to fit and they make a big difference to both your water and electricity bills because you can have quite a few showers for the cost of one bath. It is also becoming popular to fit a shower cubicle and instant shower

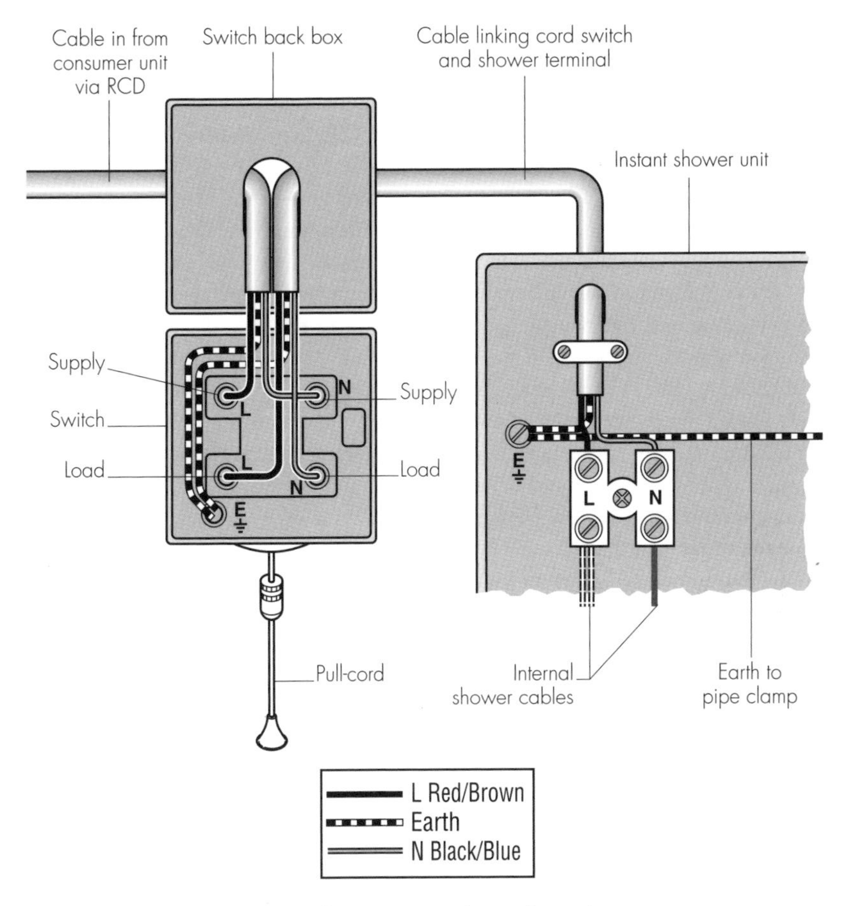

Instant shower via a ceiling pull-switch.

in the corner of a bedroom or small downstairs toilet or cloakroom.

Instant showers need a cold mains water supply, a drain and an electric supply for the integral heater in the actual shower unit. Although an instant shower is similar electrically to a cooker or an immersion heater, the difference is that you must be extra careful with the earthing in a shower circuit. The earthing must be bonded to the water pipes with earth wires and clamps. Check out "Earthing" on page 76 for detailed instructions on how to cross bond pipe work.

Tools and materials

Most electric instant showers are larger than 8 kW, so you will need a 45-amp miniature circuit breaker

(MCB) or fuse in the consumer unit or fuse box. There must be a spare slot in the consumer unit or fuse box because the MCB or fuse is allowed to supply only the shower circuit. If there isn't a spare slot, you will have to fit an additional switch fuse unit next to the existing consumer unit.

The entire installation must be protected by a residual current device (RCD), which is a special safety device that monitors the live and neutral wires simultaneously. If there is a problem and some of the electrical current is straying, the RCD instantly cuts off the electrical supply. If the consumer unit is not protected by an RCD, you should fit one in the supply cable somewhere between the consumer unit and the ceiling switch for the shower, but preferably next to the consumer unit and definitely labelled "RCD shower unit".

You will need the following tools and materials: 45-amp MCB or fuse, 10 mm² twin-core and earth cable, RCD (if needed) of at least 45 amps to suit the shower unit, double-pole ceiling switch with a pull-cord and a mechanical pop up on/off indicator, screwdriver, pliers, side cutters, wire strippers, 4 mm electrician's screwdriver, earth sleeving.

Cables installed surrounded by thermal insulation - in the loft or floor space - can have their current carrying capacity severely reduced. They need to be installed or checked by a qualified electrician.

Method

Switch off the circuit at the MCB (remove the fuse) and check that the circuit is DEAD. Run cable from the new MCB or fuse to the RCD if you have had to fit one. The RCD is simple to wire. Slip a short length of earth sleeving on to the earth core. Connect the incoming cable to SUPPLY live, neutral and earth and connect the outgoing cable to LOAD live, neutral and earth.

Then run the cable going from the RCD (or consumer unit) to the ceiling switch. This switch must be in the bathroom where the shower unit is installed and it should be labelled "Shower". Slip earth sleeving on the earth core and connect the cable to the terminals marked "SUPPLY" live, neutral and earth.

Run another length of cable from the LOAD live, neutral and earth terminals in the ceiling switch to the instant shower connection block, which will be clearly labelled, in the shower unit. The earth terminal is usually under a small metal nut or bolt. The earth should be connected to the metal parts within the shower unit. You must install cross bonding on all the pipework associated with the shower. Make sure there is a continuous earth connection from the metal parts inside the shower back through the double-pole switch, the RCD and on to the consumer unit. When you have finished the work, have an electrician check it out, especially the earthing, before using the shower.

Wiring up a power shower

Before you start

There are two types of electric shower - instant electric showers (see page 110) and power showers in which the water pressure is boosted by using an

electric pump. There are also three kinds of power shower. There is the booster type, which has a double pump in the hot and cold water supply pipes, a single impeller pump

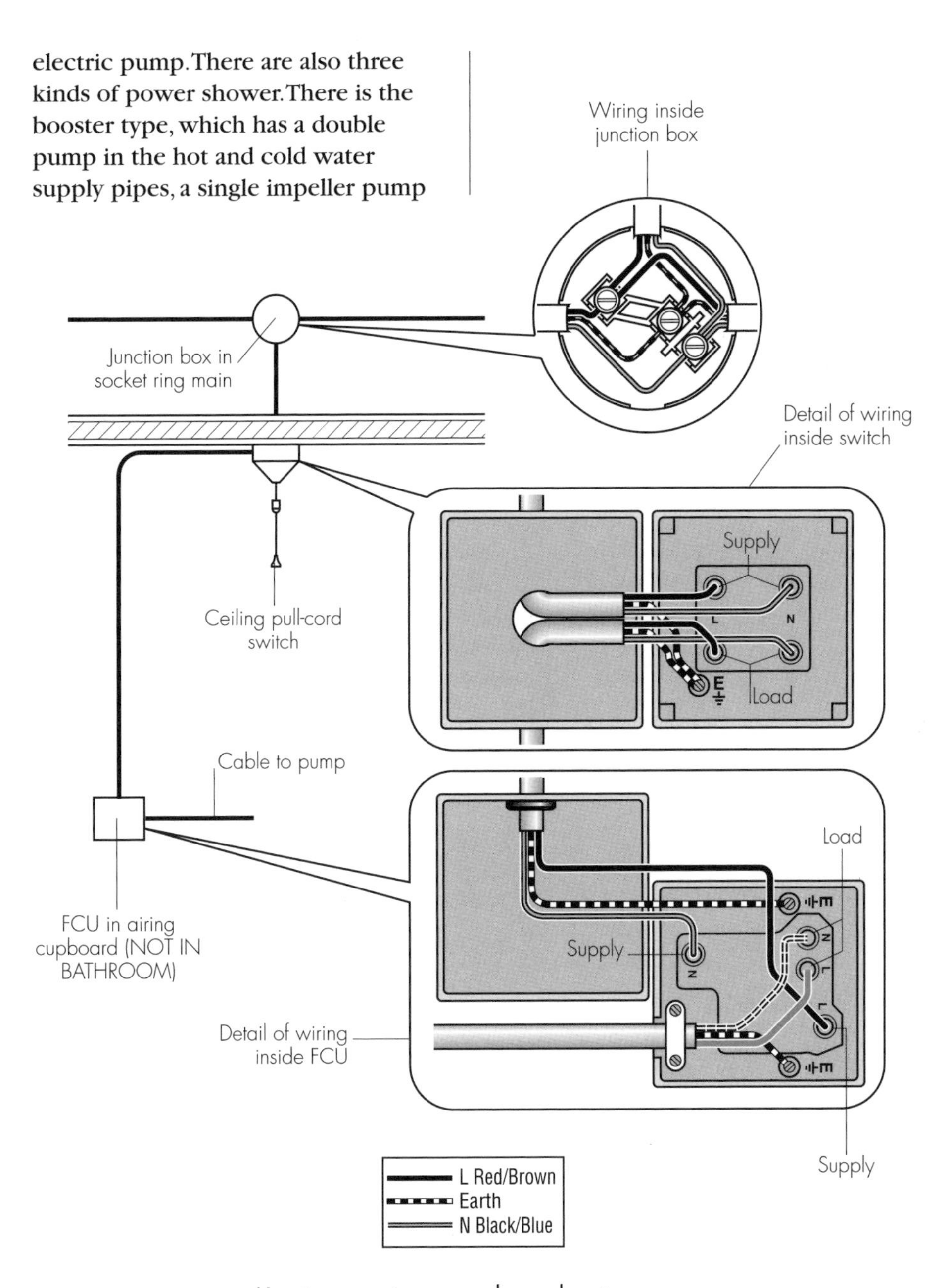

How to connect a power shower booster pump.

in between the mixer tap and the shower head, which can be fitted in nearly any home, and the type that looks like an instant shower but has an internal pump to boost the water pressure. The internal pump in the last kind is powered by the shower's internal circuit and the whole unit is wired up in the same way as an instantaneous shower. The method below therefore explains how to wire up the booster power shower.

The pump for a booster shower is normally installed either behind the bath panels or in a cupboard in the bathroom. The plumbing side of a power shower installation is quite simple. The hot and cold water pipes going to the shower mixer fitting are cut and the booster pump is installed. The pump itself usually consists of a central electric motor with an impeller pump on either side: one boosts cold water while the other boosts the hot water. The two pumps are not connected, so there is no chance of the hot and cold water mixing together until they reach the shower fitting.

Tools and materials

You will need the following: 3-terminal 20-amp junction box, screwdriver, pliers, side cutters, wire strippers, earth sleeving, 2.5 mm^2 twin-core and earth cable, 2 single backing boxes, 13-amp fused connection unit (FCU), 20-amp residual current device (RCD), flex outlet faceplate (3-terminal junction box with a cord grip and hole for the flex), 6 mm^2 earth core.

Method

Although the pump is easy to wire up, you do have to pay special attention to earthing and earth bonding.

Switch off the circuit at the MCB (remove the fuse) and check that the circuit is DEAD. Locate a socket ring circuit passing near the bathroom. It might be under the floor in an adjoining room or in the loft space above. The cable will be a 2.5 mm^2 twin-core and earth and you will find the size stamped on the outer insulation. If you are uncertain whether you have found a socket ring main cable, follow the cable back to the nearest socket. Cut the cable and fit the junction box to join the two cut ends. Run cable from the junction box to the double-pole ceiling switch. Run a cable to the FCU supply terminals from the load terminals in the switch. Fix a single backing box and connect the incoming cable to the SUPPLY live, neutral, earth terminals in the ceiling switch. Since switches in bathrooms cannot be touched, position the FCU away from harm near the power shower pump or outside the bathroom.

Fit the RCD next to the FCU, if needed, to protect you from electrocution should anything go wrong with the shower pump. Run cable from the LOAD live, neutral, earth in the FCU to the SUPPLY live, neutral, earth in the RCD.

Fit the second backing box next to the shower pump ready for the flex outlet plate. Run 2.5 mm^2 cable from the LOAD live, neutral, earth in the RCD to the flex outlet plate. The shower pump will usually be supplied with a short 3-core flex attached. Fit this flex through the hole in the outlet plate and wire it up to the live, neutral and earth terminals.

As with all electrical appliances in the bathroom, you should check the earthing very carefully and make sure the water pipes are cross-bonded. Run a single 6 mm^2 earth core from the earth terminal in the outlet plate to the nearest bonding clamp on the water pipes. Check that the hot and cold pipes are connected with an earth wire and clamps. If in any doubt about the earthing, refer to page 76 and get an electrician in to check your wiring before you take a shower.

Fitting residual current devices

Before you start

A residual current device (RCD) is a special kind of trip switch that monitors the current in the live and neutral wires. The current in both wires is always the same unless electricity is escaping from the circuit somewhere, maybe to earth or to someone who is touching a bare wire! An RCD is so sensitive to differences in current that it can cut off the electrical supply literally in a split second before anyone can be killed.

Ideally every house should have an RCD on the cables coming into the consumer unit, so that the whole house - every wire, appliance and fitting - is protected,making electrocution and house fires very nearly impossible. However, RCDs are relatively new and most houses still don't have them. In new houses the RCD may be the main switch in the consumer unit. To check if you have an RCD at the consumer unit, have a look for a switch with a small test

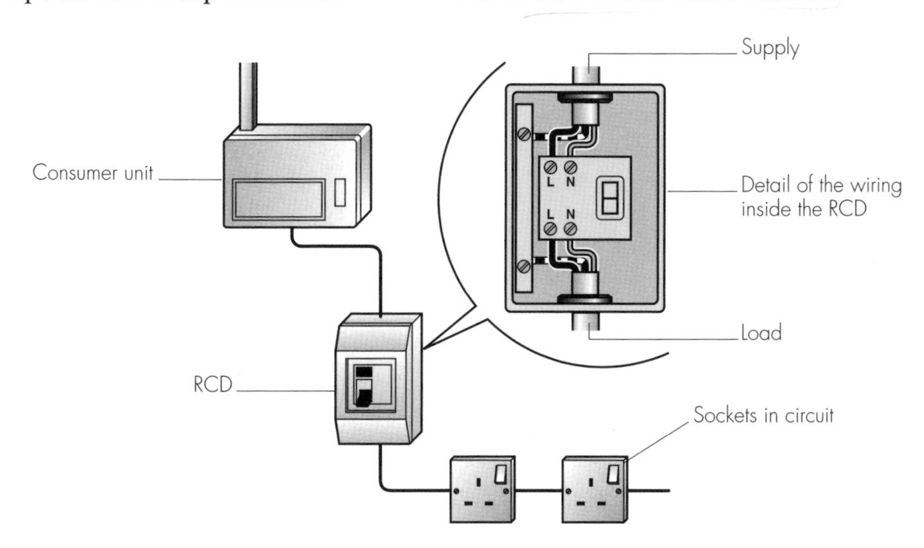

Fitting an RCD after the consumer unit.

button, which should be clearly marked "RCD".

Fitting an RCD that will protect your whole home is a job for a professional electrician from the electricity board because the cables between the electricity meter and the consumer unit will probably have to be changed. However, you might need to fit an RCD to protect a separate circuit, for example for a shower or sockets in the shed, garage or garden. Fitting this type of RCD is a DIY task and is dealt with here.

Tools and materials

RCD units are quite sensitive devices and therefore they should not be fitted in damp areas unless they are enclosed in a moisture-proof box. Inline RCDs, complete with moisture-proof boxes, are readily available. You can now buy RCDs that are designed to fit on an existing double backing box to give you two 13-amp sockets protected by an RCD. This is ideal for using with appliances such as lawnmowers, hedge trimmers and electric drills. You can locate this type of RCD where you would plug in an extension lead for working outside.

You will need the following tools and materials: flat-head screwdriver, 4 mm electrician's screwdriver, RCD that can supply an appropriate number of amps for the specific task.

Method

Switch off the circuit at the MCB (remove the fuse) and check that the circuit is DEAD. Remove the double socket from the existing backing box by taking out the two cover screws. Connect the incoming cable from the consumer unit to the SUPPLY live, neutral and earth and the cable going out to the appliance or socket to the LOAD live, neutral and earth in the RCD.

Most RCD units have a test button on the front. If this button is pushed the switch should immediately jump to the off position, demonstrating that the RCD is working.

Adding new lighting circuits

Before you start

If you build a new extension on your house you will need to run a new lighting circuit throughout the extension. Although you can extend an existing lighting circuit into the new rooms, it is better to run a completely new circuit from the consumer unit. In this way you will know that your circuit is new and up to standard - much better than tacking bits onto your existing system!

Tools and materials

Make a wiring diagram so that you can work out a materials list, including how much cable and how many junction boxes, loop-in ceiling rose fittings, light switches etc. you will need. Include the cable ducting for any cables that need to be buried in the wall to extend to light switches or wall lights. Your diagram will assist the builder in installing suitable ducting and backing boxes before the plastering gets under way. Alternatively, if the new lighting circuit is in a garage or workshop, it will be much easier to run the cable in surface-mounted trunking.

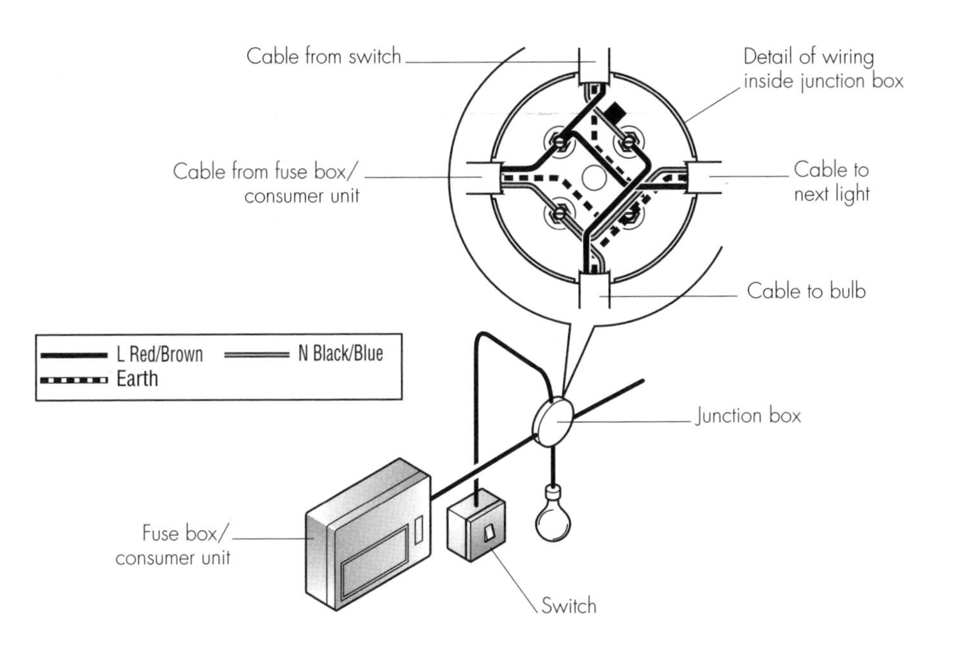

New lighting circuit using a junction box.

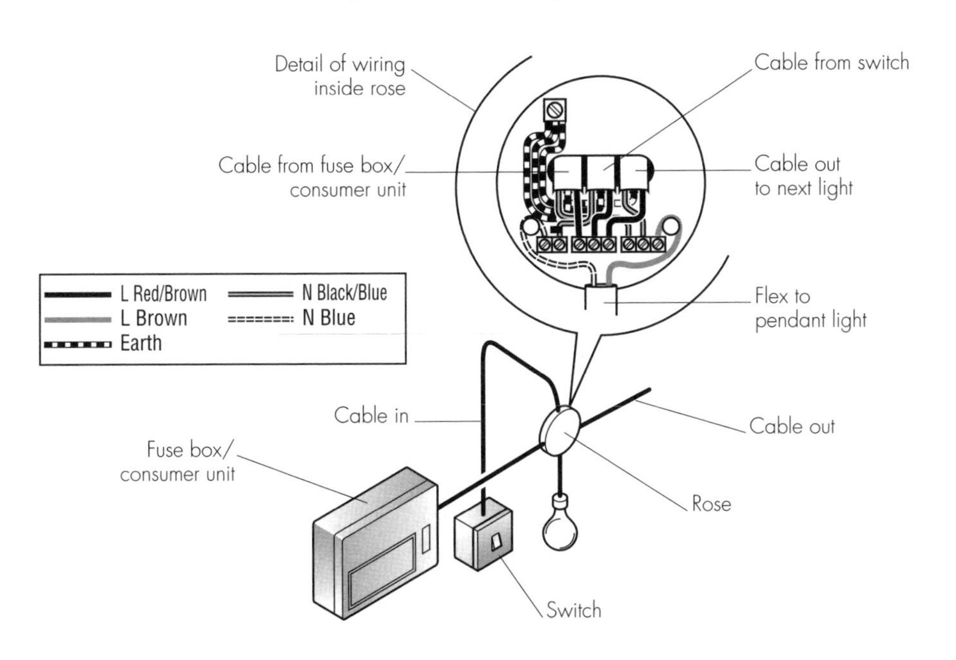

New loop-in system lighting circuit using a rose.

You will need the following: the desired number of light switches and ceiling roses or junction boxes, 6-amp miniature circuit breaker (MCB) or fuse, 1.5 mm^2 twin-core and earth cable, earth sleeving, string or wire mouse, pliers, side cutters, wire strippers, 4 mm electrician's screwdriver, flat-head screwdriver, red/brown insulation tape, 100 mm earth core, 0.5 mm^2 3-core PVC flex, 1.00 mm^2 twin-core and earth cable.

Method

Switch off the circuit at the MCB (remove the fuse) and check that the circuit is DEAD. Then add the MCB or fuse into a spare slot in the consumer unit or fuse box. Connect the red/brown live core in the cable to the top of the MCB and the neutral black/blue core to the neutral bar or terminal block in the consumer unit. Slip earth sleeving onto the earth core and connect it to the earth bar or terminal block in the consumer unit.

Run cable from the consumer unit to the first lighting position. Here you will either connect into a loop-in ceiling rose or a 4-terminal lighting junction box (see page 87). Remember to tag the black/blue core from the switch with red/brown insulation tape to show that at times it will be live, or "switched live". From the first lighting position, run the cable on to the next light fitting position and connect into the next loop-in ceiling rose or lighting junction box. Continue this process until you reach the end of the lighting circuit.

Now pull cable through the ducting or insert it in the trunking from each loop-in ceiling rose or junction box to each light switch position. See pages 80–86 for further guidance. Connect the cable to each switch faceplate, making sure that the earth wire is covered in earth sleeving. If the faceplate is a metal one, you will need to add a short earth strap, covered in earth sleeving, from the switch to the backing box. Tag the black/blue core with red/brown insulation tape.

Next add any pendant flexes that need to be connected to loop-in fittings, referring to page 46. Remember that if the pendant light fitting is made of metal it must be earthed, in which case you will need to use 0.5 mm^2 3-core PVC flex to connect the light fitting to the ceiling rose. When you have made the connections, make sure the flex is looped over the strain relief lugs, so that the weight of the light fitting is not being carried by the screw terminal connections.

You will need to run 1.00 mm^2 twin-core and earth cable to any wall light positions to 4-terminal junction boxes (see page 87). Before you connect the cable to the wall light terminal block, make sure you sleeve the bare earth core at both ends. If the wall light is metal, then also make sure that there is a good earth connection from the light fitting back to the earth in the junction box.

Adding new power circuits

Before you start

If you have had an extension built or you wish to run power to a workshop or garage you will need to add a new power circuit to the system. Although you can extend an existing power circuit by one or two sockets, this will not be sufficient for a couple of large rooms, a garage or workshop.

You will need to plan the new power circuit carefully. Look at the building plans and draw in furniture, worktops and desks. Mark where you will be using computers, televisions or power tools and make sure you have a double socket at each position. For a computer workstation you may decide to put an entire batch of sockets - perhaps four doubles - for the printer, scanner, PC, telephone, monitor, etc. A row of four sockets is a much tidier option than a single socket bristling with multiway plugs. In the case of an extension, also mark the buried ducting and backing boxes on the plans so that the builder can install these before the plastering is done. If the new power circuit is in a garage or workshop, it is an easier option to surface mount the cable trunking or round ducting.

Tools and materials

You will need the following: a 20-amp miniature circuit breaker (MCB) or fuse, a flat-head screwdriver, some 2.5 mm^2 twin-core and earth cable, pliers, side cutters, wire strippers, 4 mm electrician's screwdriver, some earth sleeving, some earth cable.

Method

Switch off the circuit at the MCB (remove the fuse) and check that the circuit is DEAD. Add the new MCB in a spare slot in the consumer unit. Loosen the lower screw on the MCB and slot it onto the rail in the back of the consumer unit, then do up the screw so that the MCB is secure. If there isn't a spare slot, you will need to fit a new switch fuse unit next to the consumer unit, referring to page 121.

Run cable from the new MCB through the ducting or trunking to the first socket position, referring to page 60. Leave about 200 mm of cable hanging out of the backing box. From there, run cable from one backing box to the next until you have completed the circuit, leaving about 200 mm of cable at each end. Run the cable from the last socket in the line all the way back to the consumer/switch fuse unit.

You are now ready to make the connections, referring to "Stripping cable" on page 54 as necessary. To connect the two ends of cable to the consumer unit, strip about 200 mm off the outer insulation. Strip 20 mm of inner insulation off the live and neutral cores. Connect the two red/brown live cores to the top of the new MCB and the two black/blue neutral cores to the neutral bar, where they could go under the same screw terminal. Slip 80 mm of earth sleeving on to the bare earth cores and then connect them to the earth bar.

To connect both cables at each socket, strip off about 100 mm of the outer insulation and cut off the excess insulation. Strip 20 mm of inner insulation off the live and neutral cores. Connect the two red/brown live cores under the live terminal and the two black/blue neutral cores under the neutral terminal. Slip 80 mm of earth sleeving on to the bare earth cores and connect

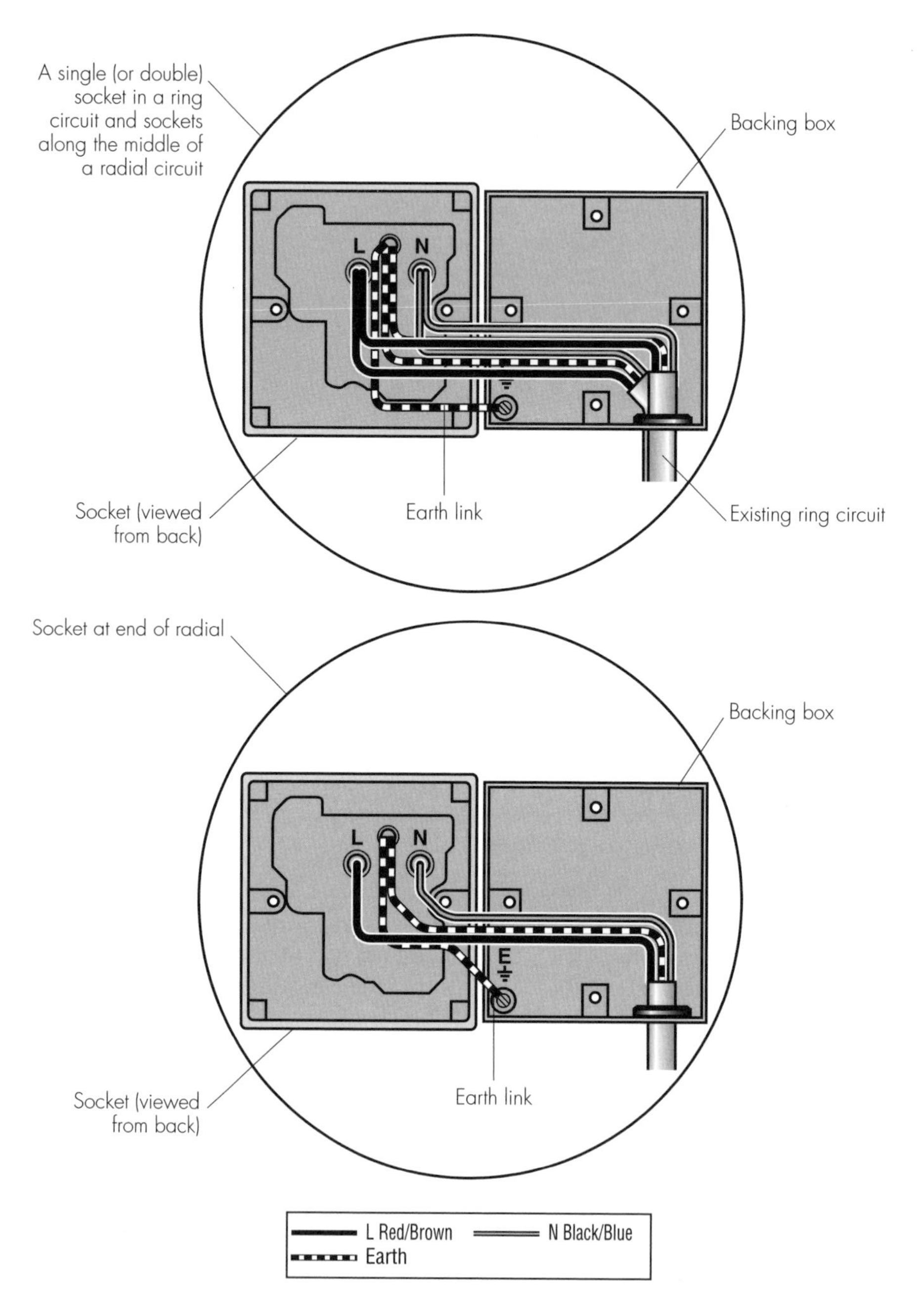

New power circuit.

them under the earth terminal. Fit a 100 mm earth strap to the backing box in each and every socket.

When you have finished the work, have a professional electrician test the circuit to make sure it is a proper ring main, continuous throughout, and that the sockets are properly earthed.

Fitting switch fuse units

Before you start

The consumer unit for your house is more than likely to have just enough miniature circuit breaker (MCB) or fuse slots to connect the circuits that were in the house when the unit was fitted. Consequently, there probably aren't any spare slots. If you have had an extension built or wish to run power to the garage or workshop, it's a good idea to have at least two spare slots fitted, one each for the lighting and power circuits. The only way to do this without replacing the entire consumer unit is to add a switch fuse unit.

A switch fuse unit looks rather like a miniature consumer unit, complete with MCBs. Some units have a double-pole master or isolation switch to disconnect the new circuits. However, a much safer option, especially if you are supplying power to a workshop, garage or bathroom where there is a risk of the wiring getting wet or cut, or where mowers and hedge trimmers will be plugged in, is a unit with an RCD earth leakage trip instead.

Before fitting a switch fuse unit, you must make sure that the total fuse ratings of the existing and the new circuits do not exceed the rating of the main electricity board fuse. You will find the main fuse near the electricity meter and distribution box, with the rating given in amps. Add up all the fuse ratings in the consumer unit and the new switch fuse unit. You cannot proceed if they exceed the rating of the main fuse, because there is a risk of blowing it. If this happens, you will have to call out the electricity board to replace the fuse – a very expensive exercise!

The new switch fuse unit should be mounted next to the existing consumer unit on the same wooden backing board. If there is not enough room, you will need to add a second backing board alongside the first. When the work is done, the electricity board must connect the large supply cable from the electricity meter distribution box to the new unit.

WARNING: NEVER tamper with the meter – this is illegal because it belongs to the electricity board.

Tools and materials

Make sure that the MCBs are the correct rating for the circuits you are going to add. A lighting circuit should have a 6-amp, power circuits a 20-amp and showers a 45-amp MCB. You will also need to make sure that the double-pole isolator or RCD in the new unit is the right size to supply all the MCBs. For example, if you have a 20-amp power circuit and a 6-amp lighting circuit you will have to fit a unit with a 30- or 32-amp double-pole switch or RCD.

You will need the following: a new switch fuse unit, flat-head screwdriver,

screws, 16 mm² single-core cable, 10 mm² single-core earth cable, MCBs of suitable rating.

Method

Secure the new switch fuse unit next to the existing consumer unit on the backing board. Connect two 16 mm² feed cables, one for live and the other for neutral, to the double-pole isolation switch or RCD in the new unit. Leave the other end of both tails for the electricity board engineer to connect to the meter distribution box.
Switch off the circuit at the MCB (remove the fuse) and check that the circuit is DEAD. In the new unit you will see the earth bar, sometimes called a "busbar", clearly marked. Run the 10 mm² earth cable, with the green+yellow sleeving, to the main earth connection on the backing board, which is usually a large screw terminal block where all the earth wires are joined up before they go to the earth rod outside your house. Make sure that all cables are neatly clipped to the backing board.

Fitting a TV aerial

Before you start

Most TV aerials are clamped as high up as possible, usually to a chimney, in order to receive the best quality signal. TV aerials are made of aluminium tubing. After a few years the tubing corrodes and becomes brittle, the wind blows and… the aerial comes down. That's one reason for having a new aerial fitted, but with the introduction of digital transmissions, known as "free view", your old aerial may need to be changed anyway because it's only suitable for receiving the analogue transmissions. However, DO NOT attempt to fit a complete new aerial yourself - it's much better to have it fitted by a professional.

However, it is possible to replace the aerial element on top of the mast yourself. Have the existing clamping bracket and mast checked by a professional. If they are in good condition, you will only need to replace the aerial element. The local DIY store should stock TV aerials suitable for use in your area.

Tools and materials

Aerials usually come flat-packed so you will need to assemble the aerial in accordance with the manufacturer's instructions. On top of the aerial you'll see a little plastic box with a clip-on cover. In the box is a small cable clamp and a screw terminal.

Aerials are wired up using coaxial cable. This is round in cross section with a single copper core at its centre surrounded by an inner honeycomb of plastic insulation. Around this inner insulation is a fine copper braid called the screen, which is then covered by an outer layer of insulation.

You will also need the following: a small adjustable spanner, a retractable craft knife, side cutters, a couple of screwdrivers, some insulation or cable ties, TV socket outlet.

Method

Once the aerial is assembled, feed the coaxial cable into the small box on top, through a tight rubber grommet, if there is one. Use the craft knife to carefully slit along about 40 mm of the outer insulation from the end of the cable, keeping your fingers well out of the way. Peel back the insulation and snip it off with side cutters. Roll the copper braid backwards so that it lies over the outer insulation. Use the craft knife to strip off about 20 mm of the honeycomb insulation so that the inner copper core is exposed.

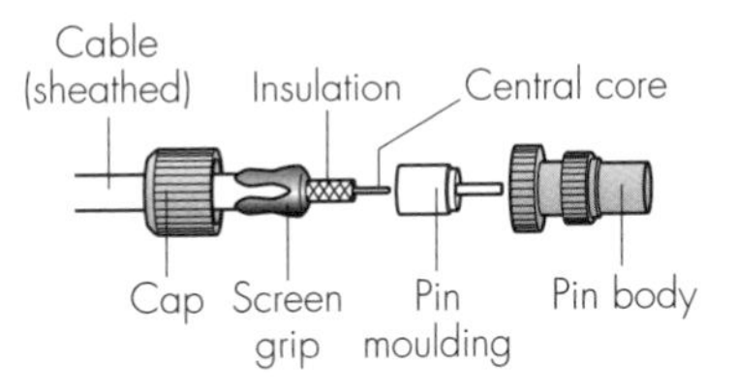

Coaxial cable and TV aerial connector – shown "exploded".

Loosen the cable clamp in the aerial box and slip the coaxial cable underneath it so that the clamp holds down the part of the cable covered by the copper braid, which makes the screen connection. Loosen the screw terminal and fit the inner copper core underneath it. Carefully refit the plastic cover and make sure the box is completely watertight.

Clamp the aerial to the top of the mast so that it faces the same way as neighbouring aerials and so that the little rods are either vertical or horizontal, again mimicking your neighbours' aerials. Attach the coaxial cable to the aerial mast with insulation tape or cable ties and run it down to the location of the TV socket outlet in your house.

WARNING: take the UTMOST CARE when working up a ladder or on the roof. Make sure the ladder is firmly positioned and held in place on the ground by a helper.

The TV socket outlet usually fits onto a single backing box. Prepare the end of the coaxial cable as you did for the connection at the aerial box. Fit the cable clamp in the backing box

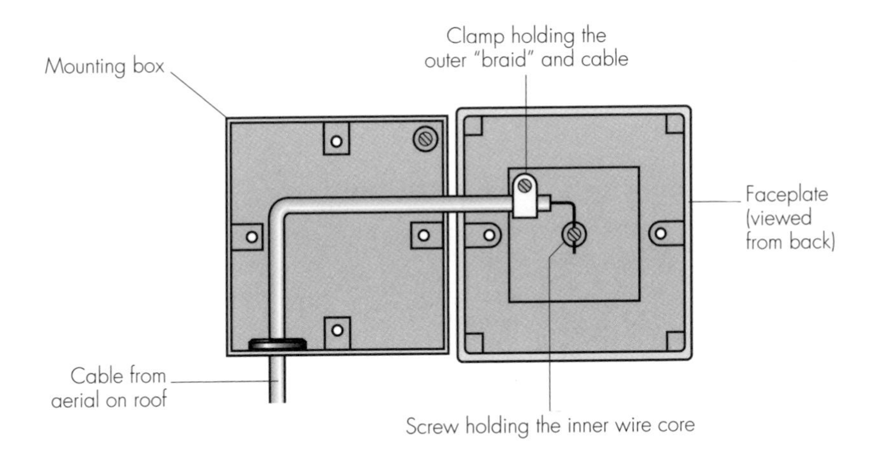

Inside a TV aerial wall socket plate.

over the part of the cable covered in copper braid and fit the central conductor under the terminal screw. Replace the TV socket outlet faceplate using the two screws supplied.

Turn on the TV and tune it in. When you have a picture, you will need to make small adjustments to the direction of the aerial. For each channel, turn the aerial slightly to the left or right until you have an interference-free, clear picture. Finally, clamp the aerial in the position that gives you the best overall picture.

Don't forget that a TV aerial is a good lightning conductor. If there is any likelihood of a storm, you should unplug your TV and VCR from the TV aerial outlet socket.

Connecting telephone extension sockets

Before you start

When the phone company connects a line to your house they will leave you with one master socket, which is usually in the room nearest to the incoming telephone line. This arrangement is not very satisfactory on two counts – most people find the position of this socket to be inconvenient and it only allows for the connection of one telephone. However, although you are not allowed to tamper with the wiring to the master socket itself, you are allowed to connect as many extension boxes to this one master socket as you like. If you create a problem in the master socket the telephone company will charge you for fixing it.

Make a plan of where you would like the new telephone extension boxes and measure up for the wiring. The wiring can be run around skirting boards, under the edges of carpets or in the walls and ceiling in the normal way. Telephone extension cable is very cheap so make sure you get a bit more than you actually need. Allow for the cable to run from the master socket, leaving about 200 mm loose for connection to each extension socket in turn, rather like a power socket

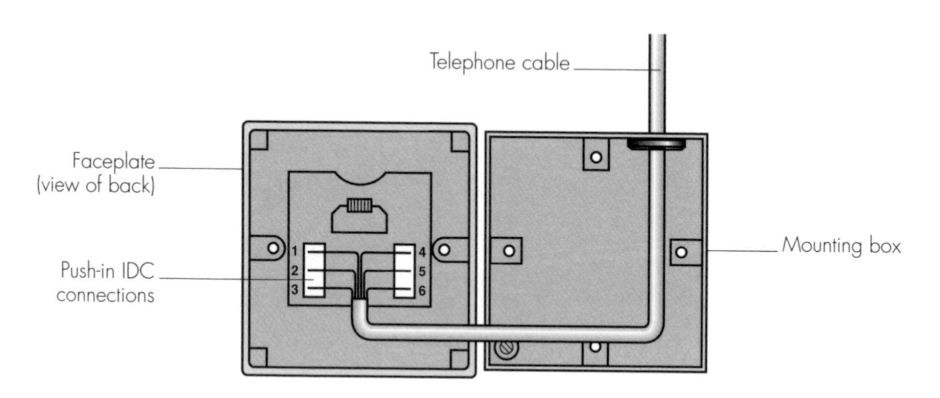

Telephone extension – last in line.

spur circuit. At each socket allow an extra 200 mm of incoming and outgoing cable for connection.

Tools and materials

The voltage used in telephone sockets is about 50 V, which is quite safe and can't electrocute you. Most modern master sockets are carefully designed so that you can connect extensions easily without affecting the original wiring. The master socket wiring is protected in the rear of the backing box, so that even while you are wiring extensions into the removable faceplate, the main wiring remains untouched and safe.

The extension boxes and cable can be bought at most DIY stores. Kits often include the special IDC insertion tool and small cable clips designed for clipping the telephone cable around the room, although these can also be bought separately.

You will also need the following: screwdriver, hammer, retractable craft knife, side cutters, IDC insertion tool, 4 mm electrician's screwdriver.

Method

Fit the extra telephone sockets around the house according to your plan, leaving the faceplates off. Run the extension cable from one socket to the next, starting from the master socket and remembering to leave enough incoming and outgoing cable at each socket for making the connections.

Unscrew the two cover screws on the master socket and pull off the cover. Remove the section that has six IDCs (insulation displacement connectors) and a plug that connects into the backing box section that belongs to the telephone company.

Carefully slit along about 100 mm of the outer insulation from the end of the extension cable with the craft knife. Snip off the excess insulation with side cutters. IDCs are designed so that you do not have to strip the insulation from each core. Simply use the plastic IDC insertion tool to push the wires into the connectors. If you look closely you can see there are small metal blades that sever the insulation and make the connection with the wires. Although most extension boxes contain six IDC connectors, labelled 1–6, most home telephone, fax and Internet equipment only uses terminals 2, 3, 4 and 5. The other connections will only be needed if you're fitting an internal telephone exchange or intercom system. Although there is a standard layout for the coloured wires and pins, it is not critical as long as you always connect the same coloured wire to the same pin number in each extension box.

Follow the same procedure to connect up the cable at each of the new sockets. The only difference is that there are two cables at each socket except the last one.

Replace the socket fronts using the screws provided and test each socket in turn by plugging in a telephone and dialling or receiving a call. Make sure that each phone on the circuit rings when someone dials in.

If you find that the telephones don't work, pull out the front of the master socket and plug one telephone into the telephone company's socket in the rear of the backing box, which should work normally. If it doesn't, then the fault is with the phone company's master socket or the telephone line. If the phone does work normally, then the problem is with your extension wiring, so check all the connections in your newly installed system.

Fitting exterior wall lights

Before you start

It's nice to have an exterior light fixed to your house wall to illuminate your path, gate or garage and they are quite easy to fit. When you arrive home on dark winter evenings you will be glad of them and of course they are an effective security measure.

Most DIY stores have a wide range of lights in all manner of materials and styles suitable for outdoor use. The lighting circuit for them is fairly simple and uses the junction box style switch connection, because you cannot use ceiling roses out of doors.

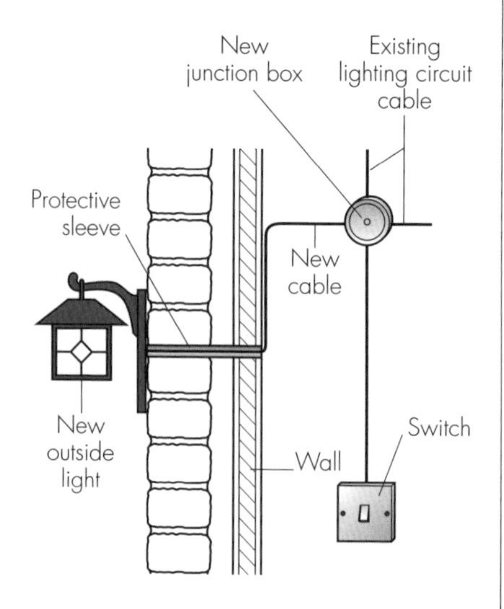

Fitting new outside light from a lighting circuit.

First determine whether you can extend an existing lighting circuit to add the outdoor lights. You will not be able to do this if the new outdoor lighting is mounted on posts or ground spikes or if the existing lighting circuit already has eight lights on it. Lighting circuits should have no more than eight lights on them, so if the existing circuit already has eight you will not be able to extend it without risking overloading the wiring. In this case you will need either to add another lighting circuit starting from a new miniature circuit breaker (MCB) or fuse in the consumer unit fuse box, or to tap into the socket ring circuit and connect a fused connection unit (FCU) to create a lighting spur. If you create a new lighting circuit, it should be protected by an inline RCD so that there is no risk of electrocution if the outside wiring is cut or damaged.

Please refer to previous sections of this book for how to strip wires (see page 54) and how to install cables in ducting (see page 60) and trunking (see page 64). This section deals with the physical installation and placement of the lights, junction boxes and switches.

Extending a lighting circuit

Locate the nearest part of the existing lighting circuit, which will probably be in a ceiling rose in the hallway. Run

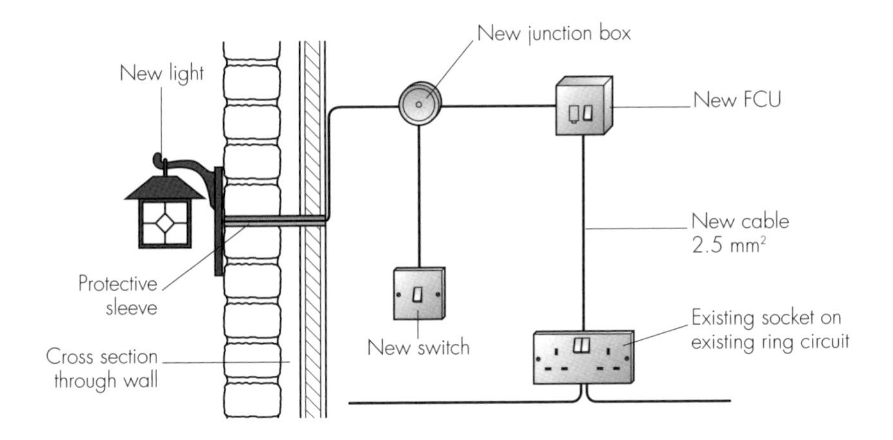

Fitting an outside light via a ring circuit.

1.00 mm² twin-core and earth cable from the ceiling rose to a convenient position, such as the ceiling or porch roof space, where you can install a 4-terminal junction box. Switch off the circuit at the MCB (remove the fuse) and check that the circuit is DEAD. The junction box should be wired exactly the same as any other junction box light switch connection and you can check the procedure on page 48.

Run 1.00 mm² cable from the junction box to the switch position. If the switch is to be placed outside, it must be a fully waterproofed IP 66 design. Otherwise the switch must be inside the house.

Run 1.00 mm² cable from the junction box to the position for the new light fitting. Fix the light fitting according to the manufacturer's instructions, making sure it is properly watertight. Connect the cable to the small connection block on the light fitting; the earth normally connects under a screw or bolt.

Running a new lighting circuit

Refer to "Adding new lighting circuits" on page 116. Make sure the new light fitting is designed for exterior use and that if you fit the switch outside it is IP 66 rated.

Connecting to a socket ring circuit

You need to locate the nearest socket on the ring circuit. Run 2.5 mm² twin-core and earth cable from the socket to an FCU, which should have a switch and 5-amp fuse, but not a red neon indicator because the unit will be on all the time. Run a 1.00 mm² twin-core and earth cable from the FCU to the junction box. Refer to the section on installing an FCU on page 95 for connecting the cables and then follow the section on extending a lighting circuit on page 86.

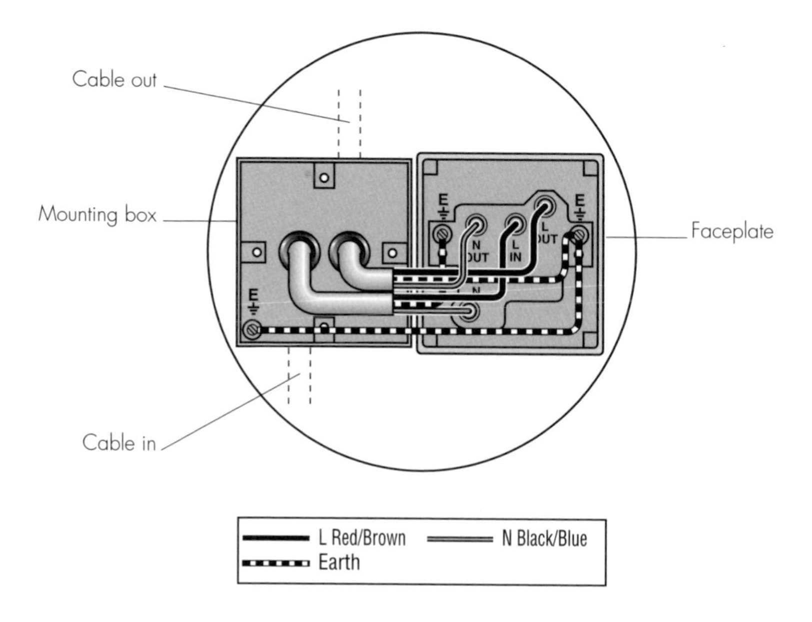

An alternative pattern of fused connection unit.

Fitting exterior low voltage lights

Before you start

Low voltage lights can look really good around your garden or in your pond, and their beauty is enhanced by the fact that it is safe to trail the cables around the garden or fence.

Tools and materials

Your local DIY store will have a range of low voltage lighting, including miniature lamp posts to mark pathways and special underwater lights to put at the bottom of your pond. There are also special bulbs to use in the lamps.

Low voltage lighting is usually 12 volts (V) and quite safe; even if the cable is cut or damaged you can't be electrocuted. It uses a transformer, which must be inside the house, to drop the voltage from 230 to 12 V.

The transformer itself only uses a minuscule amount of electricity, so it can be left on all the time with switches along the circuit for turning the lights on and off. The transformer must be installed somewhere with good ventilation, so it is usually best to put it in the loft, screwed on to a piece of plywood. Don't put the transformer in an airing cupboard or on a hollow plasterboard wall because it may overheat; although most transformers have a safety cut-out inside, it would be inconvenient if the lights went off.

Transformers will only supply a certain number of lights so check that the unit is suitable for the number of lights you want with the manufacturer. If you need more lights, you can install two transformers side by side, wired up to two separate fused connection units (FCUs).

There are also various special joints to use with low voltage cabling in a garden, available in most DIY stores. They are watertight and can be buried or left on the surface.

Method

Find a socket ring circuit near to the position for the transformer. Turn OFF the main power before you start work and then tap into a socket ring circuit by fitting a 3-terminal junction box in the cable or else connect into a socket. Run 2.5 mm^2 twin-core and earth cable from the junction box or socket to a suitably placed FCU. Transformers usually have a flex with a 3-pin plug fitted, so you could plug it straight into a socket. However, you probably won't have a socket in the loft, so it is best to connect the transformer into an FCU with a switch, red neon indicator and a flex outlet hole in the front. The FCU should have the smallest possible fuse installed. Refer to the section on installing an FCU on page 95.

Connect the cables coming in from the socket ring circuit to the terminal marked SUPPLY live, neutral and earth. Thread the flex through the hole in the front plate, secure it under the flex clamp and connect it to the LOAD live, neutral and earth. The cable from the transformer to the light fittings is usually 2.5 mm^2 twin-core.

Although the low voltage lights designed for installation in garden ponds are fully waterproof, the cable connections must always be located outside the pond itself.

Fitting exterior sockets

Before you start

If you have a garden or a garage, or a hedge that needs clipping, there will be a time when you need to use a power socket outside your house. Although you can plug in an extension lead with a portable residual current device (RCD) plug inside the house and run it out through a door or window, this is untidy and potentially dangerous. It is a much better option to fit a special waterproof socket protected by an inline RCD, on an outside wall.

There are two options for installing an outside socket. You can either mount it on an outside wall of your house or locate it somewhere less obvious in your garden.

If you must use an extension lead outside, it is most important to unroll it completely before you use it, because coiled or rolled cable or flex creates an electromagnetic field when power is being drawn through it to an appliance. This electromagnetic field could well cause the extension lead to overheat to the extent that it might melt or catch fire. If the extension lead does start to overheat, unplug it immediately, leave it until it's cool and then throw it away, since you can no longer be sure of its safety.

Make sure that there is a plug-in RCD between the extension lead and socket in your house. This is important at all times but especially if it is raining or there's a heavy dew on the ground.

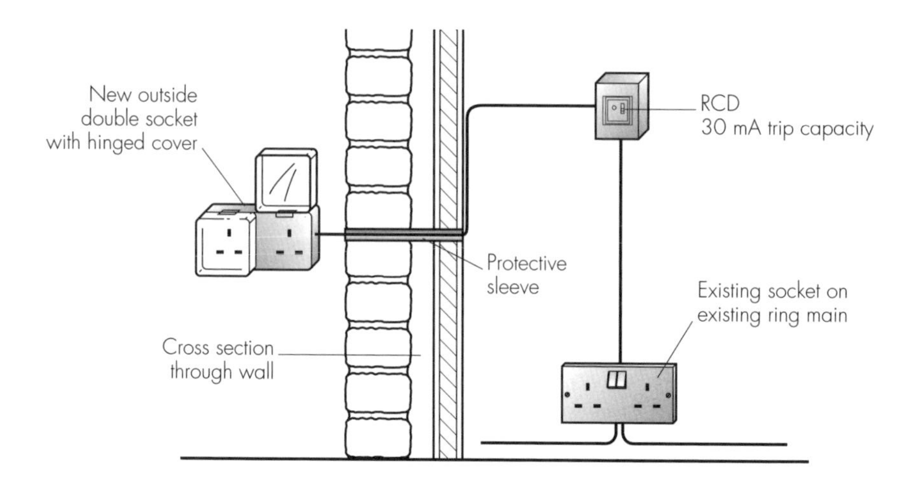

Fitting an outside socket.

A good reliable RCD can be expensive, but it will protect you from electric shock should the appliance fail or the extension lead be cut or get wet.

Tools and materials

Although there are specialist kits available for providing an outside socket, these can be very expensive, especially when you consider that such a kit usually contains no more than an inline RCD, a length of cable and an outside socket. It is much cheaper and just as easy to buy the individual components separately from an electrical wholesaler.

Waterproof sockets are available from DIY stores and electrical wholesalers. They have a hinged cover that can be lifted to plug in the appliance and then closed to seal around the flex, to give protection from damp. Although dedicated exterior sockets are expensive, don't be tempted to use an interior socket. These sockets are not designed for exterior conditions and would be extremely dangerous, especially if moisture were to gather in the base.

Mounting a socket on an outdoor wall

Switch off the circuit at the MCB (remove the fuse) and check that the circuit is DEAD. Using 2.5 mm^2 twin-core and earth cable, run a spur from a socket on your existing ring circuit to a suitable position for mounting an inline RCD, referring to pages 29 and 114. Drill a hole through the wall and insert plastic ducting to carry the cable to the outside socket. Make sure the outdoor socket is mounted correctly, paying particular attention to the manufacturer's instructions that will be supplied with the waterproofing.

Mounting a socket away from a wall

If the new outdoor socket is not going to be mounted on the outside wall of the house, but, say, on a post at the other side of the garden, it must be connected to an unused miniature circuit breaker (MCB) or fuse in the consumer unit or fuse box. First turn OFF the main power supply. Install a 16-amp MCB in the consumer unit. Working with 2.5 mm^2 twin-core and earth cable, connect an inline RCD, referring to page 114. It is best to mount the RCD on the same board as the consumer unit and it should be clearly labelled "Outside socket RCD". Connect the RCD with 2.5 mm^2 cable to the outside sockets. This cable must be protected by a plastic or steel ducting and buried at least 450 mm underground. As an alternative, you can use wire-armoured cable which can be clipped directly to the wall, run along posts or buried underground.

Running power to outbuildings

Before you start

Running power to a garage, barn or workshop for lighting and power sockets is a fairly easy DIY task.

Method

Switch off the circuit at the MCB (remove the fuse) and check that the circuit is DEAD. Install a 30-amp miniature circuit breaker (MCB) or fuse in a spare fuse slot in the consumer unit or fuse box. If the consumer unit does not have a spare fuse slot, fit a new switch fuse unit next to the existing unit, referring to page 121.

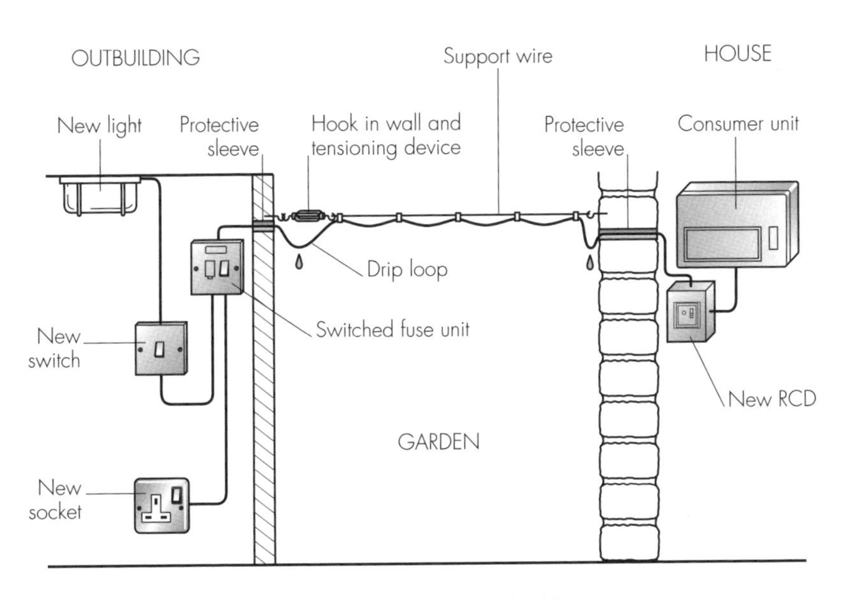

Running power to an outbuilding.

Run a 2.5 mm² twin-core and earth cable from the new 30-amp MCB to an inline RCD mounted nearby (see page 115 for how to install an RCD). If the cable run to the outbuilding is much longer than a few metres you will need to use 4 mm² twin-core and earth cable throughout the circuit.

From the inline RCD to the outbuildings you can run the cable either overhead on a support wire or underground in PVC ducting. The support wire should be heavily galvanized and fixed to eyebolts at either end, at least 3.5 m above garden level, or 5.2 m above if it crosses a road or drive. The eyebolts should be fitted either to expansion bolts plugged into the wall or to screw eyes. Use cable ties to clip the power cable to the overhead support wire. Running the cable underground is a good alternative because there is much less risk of damage to the cable. However, it will cause upheaval to your garden and paths while it is

DIAGNOSING COMMON PROBLEMS WITH INSTALLATIONS

SYMPTOM	POSSIBLE FAULT	ACTION TO TAKE
I have wired up a new light and it is always off.	The switch cable is not correctly connected.	Check that the switch cable red/brown core goes from live in the ceiling rose to the switch, and then back again to join the brown core of the flex going to the pendant.
I have wired up a new light and it is always on.	You have wired up the pendant flex incorrectly.	The blue wire goes with all the black/blue wires. The brown wire is by itself with the red/brown tagged black/blue wire from the switch. Check the ceiling rose circuit diagram on page 46.
I have put in a light and the fuse/trip goes when the switch is turned on.	The live and neutral wires to the light are being connected with the switch.	Check that the switch is correctly wired into the ceiling rose, use the ceiling rose circuit diagram and follow it carefully wire by wire.
I have put in a new lighting circuit and some of the lights don't work.	The loop circuit installed in your new system is not continuous.	Check the last working light and make sure the loop out cable is properly connected. Follow this to the next rose and check this as well.
My new dishwasher or washing machine has just been fitted and it keeps tripping out the main RCD (the whole house).	Appliances are tested in the factory and the water left inside them can cause problems if the unit is not transported the right way up. It is an earth fault in the new equipment.	Place the appliance in the middle of your room and turn the heating up all day to dry it right out. Reconnect the appliance and test it. Do not dismantle the new unit because you will invalidate the warranty.

being installed. Ideally a buried cable should be located at least 450 mm below the surface of the ground, although this distance can be somewhat reduced if it is buried under a concrete surface.

In the outbuildings, you'll need to fit a switch fuse unit that has at least two fuse ways - 5-amp for a lighting circuit and 20-amp for one or two sockets. First mount a piece of plywood board on the wall and then fix a new switch fuse unit to the board. Refer to pages 116 and 118 for detailed instructions on adding a new lighting circuit and a new power circuit. Remember that you will need to use 1.00 mm² twin-core and earth cable for the lighting and 2.5 mm² for the sockets. All the sockets and switches should be suitable for external use, even if they are to be mounted inside a shed or garage, and therefore metal clad, so that they will withstand rough treatment without sustaining any damage.

SYMPTOM	POSSIBLE FAULT	ACTION TO TAKE
My new telephone circuit works but my phones won't ring when anyone tries to make a call.	The master socket is not working, and this has a capacitor which makes all the extension telephones ring.	Replace the master socket, after carefully checking the wiring. The master socket may be the property of the phone company; check this out.
My new energy-saving bulb doesn't work.	Energy-saving bulbs won't work with a dimmer switch.	Plug the energy-saving bulb into a normal switched light and check it. Fit a normal filament (traditional) bulb in the dimmer controlled light.
The RCD supplying my lawnmower/hedge-trimmer trips every now and again.	There is an earth fault in the mower or the flex. Often a small break in the copper cores can cause this to happen.	Turn off the power, and work your way along the wire to find the damaged area where there may be a break. Fit an inline plug and socket or better still replace the lead. If the lead is OK you will need to get the mower or trimmer properly serviced.
The power cable to my workshop looks OK but is warm to the touch.	The cable size is wrong, causing the copper cores to be overloaded.	Go up at least one cable size, for example from 2.5 mm² to 4.0 mm².
One of the sockets on a double socket faceplate isn't working.	The socket has internal damage to the connection terminals.	This cannot really be fixed as the sockets are essentially disposable; replace the double socket faceplate with a good brand-name product.

CHAPTER 6

Appliance repairs

Is it under guarantee?

Most new appliances are under guarantee for 12 months, and during this period the manufacturer is generally responsible if the device malfunctions in any way. If you have misused the device in any way you may find that the warranty is invalid so take care and only use appliances in accordance with the manufacturer's instructions.

Appliance testing

Any appliance that has been repaired must be tested to make sure that is safe to use and still conforms to the manufacturer's safety standards (PAT testing).

Is it worth repairing?

Generally, if an appliance that you have purchased is at the top end of the quality available, it will be worth repairing. Appliances such as electric kettles, toasters, telephones and food processors are nearly always uneconomic to repair - manufacturers describe this as BER (beyond economic repair). Equipment like washing machines, tumble dryers and dishwashers are nearly always worth repairing, as long as the casing is not in a rusty or otherwise poor cosmetic condition. Certainly the life of an appliance like a washing machine should be a minimum of five years; with repairs and maintenance this could easily be 10 years. Look carefully at the cost of a replacement and decide whether the appliance is economic to repair.

Cautionary notes

You should remember that most appliances nowadays are carefully designed so that the manufacturer can tell whether you have taken the appliance to pieces. Manufacturers use unusual screws and fastenings as well as labels across casing joints so that they can tell whether you have been inside the appliance. If the appliance is still under warranty, do not attempt to open the casing as the manufacturer will certainly discover that you have done this and your warranty will be rendered null and void. Some appliances can be repaired quite easily with a limited toolkit: for example, dishwashers, tumble dryers and vacuum cleaners. Other appliances, such as washing machines, will need specialist tools: for example, bearing pullers to remove some sections of the washing machine. So, have a good look at the job first and make sure that you are equipped to deal with it with the tools you have available: it is obviously not economic to have to spend £200 on a tool to repair a £100 appliance!

- Be sure that it is definitely an economic option to repair rather than replace the appliance.
- Ask yourself if you can you live without the appliance for the length of time that it will take to repair it: after all you cannot do without your washing machine for a month unless you live near a laundry or have family or neighbours who are willing and able to help you out with laundry facilities.

The most common problems

Most of the problems with appliances are mechanical faults: modern solid state electronics are extremely reliable and will generally outlast the appliance. Motors, bearings, belts and brushes all have a fairly short lifespan and will be among the first things to fail in any appliance with moving parts. So, when looking for a fault it is sensible to check out the mechanical workings first.

- The flex that supplies electricity to appliances like hairdryers and electric kettles is subject to many thousands of bending loads in its lifetime. A common fault is to have a break in the cable where it is always coiled up in the same manner and a sharp bend will in the end cause a break in the copper cores. Uncoiling the flex from time to time will help prevent this happening.
- Usually the largest problem involved with fixing a mechanical appliance is being able to take the thing apart without wrecking it. Nearly all small appliances are designed with small hidden plastic clips which will break off if you don't release them carefully. It pays to study the appliance very carefully before you start work.

Do not attempt this!

There is one appliance in the home which can only be described as deadly and that is the microwave oven. A microwave oven must only be dismantled or serviced by a fully qualified engineer working in a special environment with microwave radiation detectors. The magnetron in the microwave creates the microwave energy which cooks your food. If the magnetron is not enclosed in the casing of the microwave oven, the energy will effectively be cooking you! This can result in internal organ failure or brain damage. Even a small problem like a dodgy hinge or catch on the door can allow the microwave radiation to escape. If the microwave is in any other condition than perfect, unplug it, remove the plug, cut off the lead and buy a new one.

USEFUL TIPS FOR APPLIANCE REPAIRS

- Repairing appliances mostly requires working through a process of elimination to pinpoint the exact nature of the problem. After all, most devices are extremely complicated and have hundreds of working parts. You need to narrow down the problem using logic. Draw up a list of what happened when the machine failed, and any things that the machine still does in the usual fashion. Try to narrow the problem down to a component shortlist, for example, motor or belt.
- Many appliances are put together with unusual fittings, for two reasons, mainly so that the machine is essentially "tamperproof"; the other reason is that the power tools used in the factory when building the machines work better with certain exotic screws and bolts. You can buy very cheaply a set of "security bits" which will fit in a normal hexagonal screwdriver handle: these have a full range of the various special screwdriver bits needed for modern appliances.

Washing machine repairs

Although a washing machine is fairly easy to repair, it is one of those appliances that must be sorted out quickly as there are very few of us that can survive for weeks without having access to the washing machine. Using the table below, try to diagnose the likely problem before you take the machine to pieces. Test the various functions and facilities on the machine to work out exactly where the problem lies.

Washing machines come apart in two different ways, but universally the top can be removed so that you can have a good look around to see if you can spot any likely problems. If you look at at the back of the machine, you will see two or three crosshead screws that hold the lid in position. Remove these and slide the lid forwards slightly: this will allow it to be lifted off. Looking into the washing machine you will see the drum with its concrete weights on top, an electrical clock/timer, a motor with a belt driving the drum, a soap drawer and valve arrangement and a pump located right at the bottom.

DIAGNOSE COMMON PROBLEMS

SYMPTOM	POSSIBLE FAULT	POSSIBLE REPAIR
The machine has power but it won't start the cycle.	One of the most common faults is a problem with the door sensor: the machine will not start work unless it knows the door is closed.	Open the door and remove the screws that hold the door switch, which is part of the catch, in position. Remove the switch and replace it.
The machine fills with water and makes a noise but the drum won't turn.	This can be either a failure of the heating element inside the drum or the motor or belt is not functioning.	If the motor brushes look okay and the belt is in position, it is likely there is a problem with the heating element. This is located in the bottom of the drum and there are usually three wires leading into it.
The machine overfills with water and floods the floor.	The water level sensor is not working or the tube going to it is blocked.	Remove the tube and make sure it is all clear. The level sensor is at the top of the tube; blow into the tube and you should hear a clicking noise within the level sensor if it is working. Replace the level sensor if necessary.
Water floods the floor but the machine operates normally.	The waste water outlet pipe is blocked at some point.	Call a plumber or attempt to clear the blockage yourself.

Replacing a motor

A washing machine motor is quite easy to replace with only a few tools. However, make sure you buy the right model by quoting both the machine model number and serial number when you order the spare part. You will find these numbers in the manual.

It is possible to buy a service exchange motor from a spare parts supplier. There will be a considerable saving in purchasing a service exchange unit as as you will only be paying for the replacement components in the motor, not the entire motor body. You have to take your old motor in to the supplier and swap it for the new one.

TOOLS AND MATERIALS

- Metric ring spanners up to 17 mm
- A long heavy screwdriver
- A No.2 pozi drive screwdriver
- A new motor or brushes as necessary

Basic procedure

1 Unplug the machine and disconnect the water supply.

2 Remove the lid from the machine by taking out the two or three screws at the back and sliding the lid forwards/ backwards and lifting it up. You will see the motor, usually silver in colour and connected to a pulley on the back of the drum by a belt.

3 The brushes are small black carbon blocks, which are in plastic holders either side of the motor, and you will see a small spring that pushes the blocks into each side. The brushes either unclip or unscrew and can be replaced easily. New ones will be about 20 mm long and ruined ones will be about 5 mm long.

4 You will see two or three nuts and bolts holding the motor in place. One of these bolts will be in a long slotted slide, the others will be fixing the motor like a hinge. Undo these bolts carefully and note on a sketch pad which way they are fitted and which side the nut is in each case. As the motor comes loose you will be able to swivel it inwards and lift off the belt. Unplug the electrical connector going into the motor, and there may be a clip or catch which you have to release for this.

5 Check that the new motor looks identical in every way to the old one and put it in position. Fit the three bolts loosely and lift the belt into place. For various reasons, manufacturers can change motor specifications or use a different motor supplier, so the same model can have different motors fitted. If in doubt, double-check you have a suitable replacement before opening the packaging.

6 Tighten the two hinge bolts slightly and swivel the motor out so that the belt is tensioned. You should be able to move the belt about 10 mm if you pull it halfway between the motor and the drum pulley. You may need to use the long screwdriver as a lever to hold the motor round while you do up the third bolt. Do up all the bolts firmly and replace the electrical connector and the lid. Finally, test the machine.

10 Expert Points

TOP TEN USEFUL TIPS FOR WASHING MACHINE REPAIRS:

1 HAVE A GOOD LOOK AROUND
Repairing a washing machine is all about observation; the problem is nearly always there to see if you look. When the power is off rotate the motor, drum and pump to check they will move freely without problems.

2 CHECK THE BRUSHES
The two brushes pass electricity to the motor core. They should be about 20 mm long when new and look like small black bricks. When they are worn out they are about 5 mm long.

3 CHECK THE BELT
The belt should not be perished (cracked) or have little threads hanging off the side of it. Replace it if it looks like this.

4 FRONT DOOR SEAL
If your washing machine door dribbles and leaks it is normally because the door or the seal have become dirty. It only takes a little dried soap, scale or clothing threads to break the door seal and allow leakage. Keep it wiped clean.

5 LEAKY HOSES
The hoses behind your washing machine are the most common source of leaks, because the water will invariably puddle under the machine. However, although you think the machine has a problem, this may not be the case!

6 BLOCKED PUMP
In the bottom of the washing machine there is a pump which removes the dirty water from the machine and pumps it down the drain. This pump soon becomes jammed up with fluff, hair, soap, scale and grit from your pockets! If the pump is not performing correctly take it out and clean it – usually it will survive. In some machines there is a pump filter that can easily be removed and cleaned.

7 DODGY TIMER/PROGRAMMER
If your washing machine is normally used on just one programme and starts doing something strange, try using the machine on a similar, but not the same, programme. It may be that mechanical parts in the programmer have become worn. If the machine is working properly on another programme it may be time to replace the timer/programmer.

8 WATER DRIBBLES FROM AROUND THE SOAP DRAWER
Soap will eventually cake up the soap drawer and this causes water to overflow sometimes. Simply clean out the drawer thoroughly and refit it.

9 WATER IS FOUND UNDER THE MACHINE
Check that the hoses and the back of the machine are not damaged, then have a good look at the door seal between the washing machine casing and the drum, inspecting it for cracks and splits. Also check that the hose from pump to drum is intact. Sometimes the seal at the back of the drum where the pulley is connected can leak. There is a short hose connecting the soap drawer and the drum, and this may also leak on occasion.

10 DOOR SWITCH
The door switch on a washing machine can be at the root of several problems. The safety sensor inside it can stop the machine from working and may also stop the door from opening. Check its operation carefully and replace it if necessary. It is not an expensive part and will provide a quick solution.

Tumble dryer repairs

A tumble dryer is quite a simple device - it contains only the timer, drum, heater and safety sensors. If the tumble dryer does not heat up then one of the elements may be broken. If the drum will not turn, either the motor, brushes or belt may be at fault. The heating elements are generally at the back of the drum and look like giant kettle elements. Usually there are two elements: one that comes on for lower heat and the other that comes on as well for high heat. If the machine is not heating up fully, it may be that only one element has failed. As in a washing machine, the brushes in the motor gradually wear away: these are the little carbon bricks on either side of the motor. As with washing machines the brushes should be about 20 mm long when brand-new but will wear to about 5 mm before the machine stops working.

The timers on tumble dryers rarely go wrong as they are so simple. There is a safety device to stop the door being opened, either a switch in the door catch or a pull-wire which allows the door to open as long as the motor is not turning.

Dishwasher repairs

Dishwashers are fairly easy to repair as they have actually only a few components. There is a pump in the bottom of the machine which cycles the water from the base up to the rotary sprayers, there is a heating element and a programmer/timer and a discharge pump. The water is allowed into the machine by an electrically opened valve where the water supply hose connects to the dishwasher. You need to have a play with the machine to work out which item is not working. The machine is fairly easy to take to pieces but you will find large amounts of insulation inside the casing as dishwashers do get quite hot.

Kitchen gadgets

Kitchen gadgets such as electric tin openers, food processors etc. all have a few basic things in common: a switch, a motor and a gearbox or drive system. Many are carefully designed to stop you opening the casings! Small screws are hidden under labels or rubber feet - if it won't come apart have a look under the labels and feet just in case. For speed of assembly in the factory, some units use clips which lock into place and hold the casings together; these can be difficult to release, or even to find in the first place.

Most gadgets of this type are not worth repairing unless you are simply curious about why they won't work and have the time to spare to investigate! Start at the plug and flex and check that they are okay, and work your way to the switch and make sure that is also functioning correctly. Most motors will have brushes of some kind which wear away after a time and can be replaced if the machine is still manufactured. Often small machines have plastic gearbox or drive components, which will wear out or get smashed up over time. It is not really worth replacing these unless you take it as a challenge!

Cooker repairs

Electric cookers may appear to be quite complicated pieces of equipment, but in reality they are made up simply of six switches and six heating elements. Of course before you start work on a cooker you must ensure that it is fully disconnected from the power supply – this includes removing the cable from the wall after turning off the main switch. Usually with a cooker it is only one ring or the oven which is not working, so it is easy to track down the problem either to the switch or to the element itself. The switches are generally long-lasting since they have inside them a small heating element and metallic strip which controls the temperature of the ring. Over time the switch may fail, and you can tell if this is the case because it will generally work when set to "maximum" but will not properly operate at any other setting. If your cooker is a brand-name product you will be able to get a replacement switch fairly easily. The elements or rings are removed by lifting off the cooker top or removing panels within the oven. The element itself is usually screwed or bolted into place and there will be two wires connected to it.

As you cannot easily do without a cooker for very long it is best to obtain the switch or element before you start work on the repairs. If you look at the back of the cooker you will find the manufacturer's model number and serial number. You will need to have these to hand when ordering spare parts.

Microwave oven repairs

As stated elsewhere in this book, a microwave oven is an extremely dangerous device and should not be dismantled under any circumstances. The magnetron inside the microwave oven produces microwave radiation which can cause you serious damage or even death. Never remove the covers from the microwave and never operate the microwave unless the door fits perfectly. If in doubt take your microwave to a repair centre and have it tested to make sure no microwave radiation is escaping. There is only one thing you can replace easily and safely in a microwave oven and that is the small lamp that lights up the inside when in use. This is accessed from inside the microwave compartment. Where the rotating dish is, you will see a small transparent cover; the lightbulb is behind this. Use a screwdriver to pop off the panel or unscrew it as necessary, and replace the bulb with one of a similar style and power.

Vacuum cleaner repairs

As you can see from looking around the shops there are dozens of types of vacuum cleaner; however, they are all remarkably similar in design. The vacuum cleaner has a plug and flex connected to a switch; from here a short cable goes internally to the motor. The motor has two carbon brushes which will gradually wear away over time – these conduct the power to the motor core. The motor itself will have a fan on one end of it built into a casing. The simplest way to repair the unit is to have a good look at all these components and check through the system logically. Some upright vacuum cleaners have drive belts and rollers underneath which can be blocked with hair, threads or kids' toys and make some horrendous noises or even burning rubber smells. Unplug the cleaner, turn it over and check this out first!

Small toys can be pulled out using pliers and hair can be removed by carefully cutting with a pair of fine scissors (do not nick the belt). At each end of the brush roller you will see some kind of pivot or bearing. These get jammed up easily, so have a look here first for any obstructions.

Diagnosing the problem

The scenario is that you have plugged in a traditional, good quality vacuum cleaner and it doesn't work. Everything looks fine, there's no sparking, smell, signs of burning or anything unusual, apart from the fact that it doesn't work. Note that a cheap vacuum cleaner is probably not worth repairing.

TOOLS AND MATERIALS

- 4 mm electrician's screwdriver
- A torch
- 13-amp fuse
- Wire strippers
- Electrical long-nose pliers
- Multimeter
- Replacement flex to match your cleaner

Basic procedure

1. Switch off the vacuum cleaner, remove the plug from the socket and have a good, long visual inspection of the whole cleaner - the plug, the flex, the point where the flex enters the casing and the casing itself. Ask yourself if anything looks unusual, cracked, loose, discoloured or misshapen, or if the flex looks fractured or is unusually limp anywhere along its length.
2. If the cleaner looks fine, find another appliance that you know is working and plug it into the same socket. If the test appliance works, then you can be certain that the cleaner is at fault. If the test appliance doesn't work in that socket, refer to "Emergency repairs to a dead power circuit" on page 155.
3. Plug the cleaner back into the socket and have another attempt at vacuuming just to make sure that you didn't do something strange the first time around.
4. If the cleaner still doesn't work, remove the plug from the socket. Open the plug and make sure that all the connections are good and tight at the terminals. Make sure that the terminal pins are tight. Either replace the fuse with a new one or check the existing fuse with the multimeter. Close the plug up, insert it into the socket again and switch the cleaner back on.
5. If the cleaner still doesn't work, pull out the plug and open the casing of the cleaner at the point where the flex enters. Follow the flex into the casing to the terminal block where

the flex is gripped and the cores are screwed to the terminals.

6 Tighten up the terminals and generally check the flex cores. Then replace the casing, push the plug into the socket and switch on the cleaner.
7 If the cleaner still doesn't work, pull out the plug, open up the casing and remove the flex, complete with the plug. Take the plug off the flex. Use the multimeter to test the flex cores. If you have doubts about the flex and you suspect that one or other of the cores is stretched or partially broken, then replace the whole length of flex. Ideally, you need a pair of helping hands, so that you can do the testing while your helper stretches and bends the flex.
8 Then refit the flex to the cleaner at one end and the plug at the other. Make sure that all the terminal contacts are tight, then plug in and switch on the power. If the cleaner still doesn't work, it is best to take it to a dedicated service engineer for a general repair and overhaul.

TV and VCR repairs

TVs can still be dangerous even when the power supply has been removed because of high voltage from inductive and capacitive loads.

Televisions and video recorders leave you in a bit of a dilemma regarding repairs. TVs do not really have any user-serviceable parts, unless a button has fallen off the front or something! VCRs are now so cheap that it is scarcely worth removing the top to have a look, and definitely not worth having a professional look at it. Limit your television repairs to replacing the flex (maybe damaged by the dog or the kids!), plug or small plastic doors that snap off the front.

VCRs are a little more repairable, if you have the time and the inclination. Mostly the VCR will pack up due to a dodgy picture; this is often from dust and debris in the reading head which looks like a silver, polished, spinning drum on an angle in the machine. This can be cleaned with a "lint free" piece of material, for example a ribbon. You can purchase a tape head cleaning cassette, which may be worth a go. It is best used on fast forward play as the cleaning tape passes swiftly over the head. There are a number of small rubber belts in the machine which drive the two tape spools around. Sometimes these bands break, but if your machine is a brand-name device you may be able to get a replacement.

DVD and CD player repairs

DVD and CD players all work using a laser beam which is fired onto the disc: the reflection of this beam is read and processed into music or video. This laser is not the same as a laser pointer used at the office, it can damage or even blind you if it catches you in the eye. Never plug-in and operate these machines with the covers removed, it is just too risky. If your DVD or CD player starts to play up, skip or fail to recognize discs, it is nearly always because the small lens covering the laser has become dusty, or dirty. You would be surprised how quickly the lens will become sooty in a smoker's living room for example. You can buy a commercial disc cleaning kit that may work but, in severe cases, you will need to dismantle the machine, after unplugging it of course, and clean the lens yourself. The best tool for doing this is a cotton bud or "Q-tip". Never use any chemicals on the cotton bud, just slightly damp it with water. Only touch the lens very lightly as it is easy to damage the small support structure which moves the lens to focus it. You will usually find this will fix the unit.

Computer repairs

Computers can still be dangerous even when the power supply has been removed because of high voltage from inductive and capacitive loads.

PCs or home computers are actually very easy to repair when there is a problem with the hardware. All of the components within the computer are easily available on the high street or by mail order and are quite universal with regard to fitting within the casing. The replaceable parts within a computer are the power supply, fans, casing and switches, floppy disk drive, hard disk drive, CD player and component boards. The computer itself can be fully dismantled with a crosshead screwdriver on your kitchen table without much problem. Make sure that you have made back-ups of your data before you take the computer to pieces! Remove the casing and you will see the different components. The power supply is a grey steel box screwed to the back of the casing where the mains supply cable goes in. A few screws can be removed and this unit can be unplugged from the main circuit board and replaced. The disk drives and CD drive are slotted into the front of the computer and held in place with four screws each. The screws usually go in from the sides. Remove the two connectors from the back of the drives, one wide ribbon cable and a small cable with four wires, and slide the drive out of its bay. Then replace the components as necessary.

Power tool repairs

Power tools only have a few electrical components, the plug and flex supplying the power, the switch and the motor. The plug and flex are easy to check and if either are damaged they should be replaced - especially if the insulation is damaged in any way. The motor itself will have two small carbon brushes, one on each side of the back of the motor. These brushes wear away over time and are the primary cause of power tool failure. If the power tool is a brand-name model, you will be able to obtain new brushes by quoting the model and serial number to the spare parts retailer. On good-quality power tools it is well worth replacing a broken trigger switch or brushes as these will be very cheap.

Jigsaws and drills frequently suffer the same problem after a number of years of use: the bearing where the blade or drill chuck enters the tool wears and becomes hot and wobbly in use. If you have bought good-quality tools, you will be able to buy individual spare parts. Consult the handbook supplied for the part number of the item you require; it can usually be ordered over the Internet or by telephone from a dealer. Check out the cost of a new tool after you have priced the spare parts - these things get cheaper all the time, so it may be simpler to buy a new unit.

CHAPTER 7

Emergencies

A QUICK GUIDE TO THE TOPICS COVERED IN THIS CHAPTER

What is an emergency?

What constitutes an emergency repair has more to do with your situation, needs and skill level than anything else. One man's emergency repair is another man's ordinary repair.

In the context of electricity, an emergency might be thought of as a situation in which there is a danger of accident to people or property. So, for example, a complete electrical shutdown on a cold winter's night is unpleasant and uncomfortable, but it cannot be compared to an electrical appliance that is on fire. On the other hand, that same cold winter's night might be a critical time for the sick, the elderly or young children, or for anyone with livestock that require milking. Then it becomes an emergency. Much the same goes for something like a faulty flex on a room heater. For some people this would be a nuisance, while for others it would be an emergency. Whatever your sense of emergency, in this section of the book you will find quick remedies for everything from a power failure to extending a flex.

If you are still a raw beginner at dealing with electricty, then a serious emergency such as smoldering wiring or a consumer box on fire is not really the time for you to put your untried

DIY skills to the test. There is every reason why you should take quick action to sort out a problem before it gets worse - so you should switch the main power supply off at the mains. If the situation is potentially life-threatening and running out of control, bring in a professional.

10 Expert Points

QUESTIONS AND ANSWERS ON COMMON EMERGENCIES:

1 WHY IS AN APPLIANCE SPARKING?

What should I do if an appliance such as the electric kettle sparks or starts burning? Go to the consumer unit or fuse box, turn OFF the main power supply, then go back to the appliance and pull out the plug. If the kettle is on fire, smother the flames with a dry powder fire extinguisher (if you have one) and cover it with a damp rug or blanket. Then carefully remove it from the premises. NEVER throw water on an electrical appliance.

2 WHY HAVE THE LIGHTS GONE OFF?

What should I do if it is dark and the lights fail? If everything appears to be safe and sound, and there are no smells of burning or sparking light switches, wait for a short while just to make sure that there are no real dangers. Then go to bed and wait for daylight. What you MUST NOT do is to start wandering around with a candle, climbing up ladders or crawling through the loft in the dark.

3 WHY ARE PLUGS AND/OR SOCKETS OVERHEATING?

What should I do if a plug and socket become hot while the socket is in use? First turn OFF the main power supply in the consumer unit or fuse box. Then pull out the plug and wait until the plug and the socket have cooled down. Open the plug up and make sure that all the terminal connections are good and tight – the earth, live and neutral connections, and the fuse. Take the faceplate off the socket, check that all the connections are tight and make sure that the cable cores are not rubbing on the backing box. Make doubly sure that all the cores go to the correct terminals. If either the plug or the socket are in any way damaged or loose, then replace them.

4 WHAT SHOULD I DO WITH A LEAD PENCIL IN A SOCKET?

My toddler has pushed a pencil into a socket and it's stuck – what should I do? Pencil leads are very good electrical conductors and very dangerous! First remove the child to another area or room, so that they are safe and you can work uninterrupted. Then turn OFF the main power supply in the consumer unit or fuse box. Go back to the socket and remove the pencil. It shouldn't be possible to poke a pencil in a socket. The fact that it has happened suggests that the socket is either very old or damaged. Remove the front of the socket and make sure that it's intact. If you are at all concerned about the state of the socket, then replace it with a childproof type. Now start teaching your child about the dangers of electricity and consider fitting safety covers for sockets.

5 WHY IS A FLEX OVERHEATING?

I have put a long flex on the fridge so that I can site it on the other side of the kitchen, with the flex clipped in place around the skirting, but why is the flex warm? An overlong flex will soon overheat, but it also sounds as if you have fitted a flex that is too small for the job. Change the flex to 2.5 mm^2 twin-core and earth, and consider extending the circuit so that you have more sockets.

WHY SHOULD I UNROLL AN EXTENSION CABLE?

I have a very long extension cable on a reel so that I can use the lawn mower at the end of our garden. A neighbour tells me that I should always unroll the full length even when working close to the house. Why?
No matter where you are working in the garden, even if you only want to use a couple of metres of flex, you must unroll the flex from the drum. If you don't, the flex could overheat and melt.

WHAT SHOULD I DO ABOUT A MISSING LEAD SEAL?

Should I be worried that the lead seal on the underside of the electricity meter in my new flat is missing? Everything on the other side of the meter, including the meter, belongs to the electrical board and if they see the missing seal they might think that you are stealing electricity. You must phone them immediately and report the problem giving dates and times. Give them a clear meter reading. Make sure that you get a name and a reference number from the person you speak to.

8 HOW DO I REPAIR A DAMAGED CABLE?

While putting a picture on the wall, I drilled into the wall and hit a cable. There was a flash and bang, and the power failed. What should I do now? Take time to recover from the shock and then switch off the circuit at the MCB (remove the fuse) and check that the circuit is DEAD.Take a hammer and cold chisel, or an old wood chisel, and carefully remove the plaster from around the drilled hole. Once you have found the cable, create a small hole so that there is room to work. Carefully cut through the plastic ducting to make a hole in it and gently ease the damaged cable though. Cut through the cable at the point of damage, so that you have two ends, and tie one very long piece of strong string to both ends of the cable.

Open the socket below the problem and remove the front plate. Gently ease the cable down and out until you have one end of the string. This done, go up into the loft or ceiling space above the problem and drag the other half of the cable through so that you have the other end of the string. Tie the string to a new length of cable. Back downstairs, pull the string to bring the cable down to the socket end.

Then connect the ends of the new cable to the socket terminals and the appropriate loft junction box in exactly the same way as the old one was connected. Finally, cover the hole in the ducting with a piece of plastic cut from something like a milk carton and then use plaster filler to restore the damage to the plaster.

9 THE DOG'S CHEWED THE TV FLEX

I've just caught our dog chewing the TV flex. It hasn't gone through the insulation and the TV still works. What should I do?
Bind the damage with insulating tape and replace the flex at the first opportunity.

10 WATER AND COMPUTERS

Water has dripped on to my computer, so what should I do?
Do NOT switch on the computer. Unplug it, remove the back panel and use a hairdryer on the lowest setting to dry out the interior of the cabinet until you are sure that it is all absolutely dry. Leave the computer in a warm room for 24 hours, then replace the back panel and switch on the computer.

Method and procedures in an emergency

Electricity is a potential killer! One ill-considered action can be a catastrophic mistake.This doesn't mean that you have to shiver and shake at the very notion of touching a socket or appliance; it simply means that you always have to follow through the same series of cool and methodical procedures before you start work. You won't go wrong if you let the Expert Points below be your guide. But don't wait until you're panicking about an electrical emergency before reading this section. Do the reading first!

10 Expert Points

SAFETY PROCEDURES THAT COULD SAVE YOUR LIFE:

1 NEVER DRINK AND...
NEVER work under the influence of drink, drugs or strong medication.

2 BE FIT FOR THE JOB
Make sure that you are physically fit. If you want to undertake a challenging DIY task, but have doubts about your physical fitness, then ask your doctor.

3 MAKE SURE THERE'S HELP AROUND
If you are elderly, not very stable on your feet or given to dizzy turns, it's a good idea to ask a friend or partner for help – to hold the torch and to pass you the tools. If you have to work alone in the cellar or loft, then tell a neighbour what you are doing. If possible carry a mobile phone.

4 TURN OFF THE MAIN POWER SUPPLY
Always start work by switching off the circuit at the MCB (removing the fuse) and checking that the circuit is DEAD before carrying out any work. If you want more evidence that the power is off, then it's a good idea to plug a radio or lamp into a socket. If you still have doubts test it out with a multimeter. If you are still unsure, call in a professional.

5 SAFETY WITH AN RCD
If you want to make changes to your electrics over a long period and be doubly sure that you can work in safety, it's a good idea to get a qualified electrician to fit a residual current device (RCD) first. You still need to follow all the safety procedures, but if you do make a mistake the RCD will save you from physical harm, by cutting the power off in a split second.

6 TAKE YOUR TIME
It is easy to make mistakes if you are trying to work in a rush or if you are over anxious. Quietly consider all the options and take your time.

7 BE DOUBLY SURE
NEVER take things for granted. You may have just pulled out a fuse or turned off a switch, but you must still make tests before you start touching cable cores.

8 TELL OTHERS WHAT YOU'RE DOING
Accidents happen when one person is working on a circuit and another person throws a switch or replaces a fuse. If there are other people living in the house, then make sure that they know what you are doing. If you have doubts, then remove any appropriate fuses and put them in your pocket until you have finished the work.

9 CHECK IF THE PROBLEM IS WIDESPREAD
Accidents can happen when there is a power failure outside the house and the homeowner assumes that the problem is inside the house. If the power fails, then start by asking your neighbours if they are having problems too.

10 GET PROFESSIONAL CHECKS DONE
If you have any doubts at all about the standard of your work, ask a trusted electrician to check it out.

Dealing with electric shock

Electric shock can kill! If the victim stops breathing, he or she can die. If you are not the only person on the scene, call or send someone to phone for an ambulance straight away.

If you see someone working on electrics and they collapse suddenly, you need to know what to do. Certainly you need to act fast, because precious minutes can make the difference in saving a life. However, in addition to speed, you need to act with cool, methodical calm.

The following steps cater for the worst case scenario. The victim has received a severe shock and fallen from the top of a ladder. They are now on the floor and still gripping the socket, wire, or appliance. Their breathing seems to have stopped. The steps are set out in sequence and it's important that you follow through from start to finish in order.

Step 1: Don't touch!

Do NOT touch the victim.

Step 1: Remove the source of the current

Go to the main consumer unit or fuse box and turn OFF all the power. If this isn't possible, use a wooden or plastic implement, such as a broom or walking stick, to knock the electrical source out of the victim's hand. If this isn't possible, then pull out the plug, protecting yourself by wearing rubber gloves if there is time to find some. Only as a very LAST resort and if everything else fails, attempt to pull the victim free by gripping a piece of loose clothing. Do NOT touch the victim's body until they are no longer in contact with the source of the current.

Step 3: Position the victim

Once the victim is no longer in contact with the current source and the power is off, gently ease them over so that their head is slightly tilted back with the tongue lolling free so that it does not block the airway.

Step 4: Give the kiss of life

If the victim still isn't breathing, pinch their nostrils closed, set your mouth completely over theirs, gently blow until their chest rises and then remove your mouth. Repeat this procedure at about 5-6 second intervals.

Step 5: Phone for an ambulance

After you have given 10 or 12 breaths and if there is no one else available to do this, phone for an ambulance and explain what has happened.

Step 6: Call for help

If you can't get to a phone, then go into the street and shout for help.

Step 7: Continue the kiss of life

Continue giving the kiss of life until the victim starts to breathe or until the ambulance arrives.

Step 8: Use the recovery position

Once the victim has started to breathe, ease them over on to their side with their head resting sideways on one extended arm. Make sure that their tongue is not blocking their airway.

Step 9: Keep the victim warm

Cover the victim with blankets or a coat. Do NOT try to sit them up or give them anything such as a tot of rum, a cigarette or a cup of tea.

Step 10: Unblock the airway

If the victim vomits, wipe away the vomit and use your finger to make sure that nothing, including their tongue, blocks their airway.

Using a multimeter to diagnose the problem

A multimeter is one of the most useful electrical diagnostic tools. However, it must be treated with caution as there is a severe danger of electric shock when testing for the presence of mains 230 volts (V). It is NOT recommended that you use a multimeter for testing for mains voltage, unless you have a very good working knowledge of electrical theory and practice.

Any voltage test probes should have no more than 1 mm of metal tip exposed, must be fused and must have finger guards.

For use around the house, a multimeter only really needs two functions. First, it needs to be able to test for 230 volts AC, whereas most small multimeters are CAT II, which will only test up to 600 volts safely. The maximum voltage rating is printed on the front of a multimeter. Second, the other very useful function of a multimeter is a continuity test so it should have a continuity buzzer mode. This option allows you to test fuses, light bulbs, cables and flexes. When you buy a multimeter, make sure that you read the manufacturer's instructions very carefully and have a good understanding of what you are trying to test for, otherwise it is easy to toast the multimeter and/or yourself!

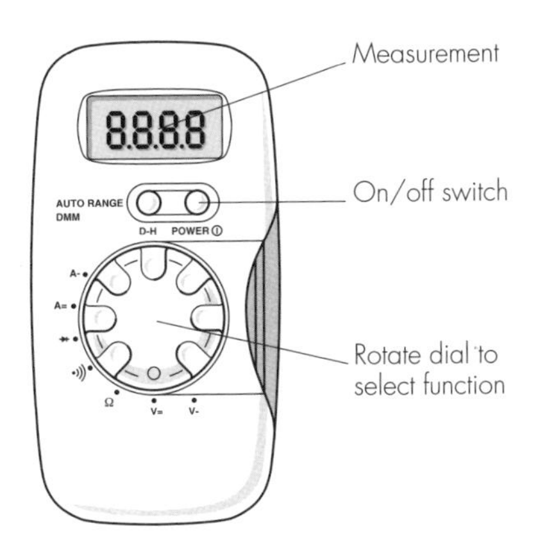

A typical multimeter.

Tracing mains voltage

A multimeter is a superb tool that can trace faults in all manner of materials and appliances - everything from extension lead breaks, plugs and fuses, through to hairdryers and appliance circuits. You can use a multimeter to show you where the power has reached, or has failed to reach, as the case may be.

WARNING: It is DANGEROUS to work on LIVE circuits. The EAW regulations state that live working is to be discouraged. Only qualified, competent electricians should work live and only then when they have put in place all the required safety precautions. Make sure that the appliance is plugged into a sensitive residual current device (RCD). Do not touch any metal parts, including the casing, wires, screws, heating elements or the metal probes on the ends of the multimeter test leads.

To measure AC volts, set the dial on the multimeter to the "V~" symbol. If the two test leads are now put onto a live and neutral terminal, the multimeter should show between 230 and 250 volts AC, depending on the electrical supply in your local area. Take care that the multimeter is not displaying 230 mV, as this reading means 230 thousandths of a volt, which could be just stray current in the wiring.

To use the multimeter to find a fault, first plug in the appliance. Then start at the plug and test for voltage there. This done, work your way towards the appliance, testing the cable, the switches, etc. as you proceed. Remember, everything is live and dangerous, so TAKE CARE!

Testing a fuse, bulb or cable

To test a fuse, bulb, flex or cable the item being tested must NOT be connected to the electrical supply. The continuity test is made by a small battery inside the multimeter, so it will be safe to touch the wiring and multimeter probes at any time.

WARNING: Before testing, always check that NO part of the item being tested is connected to the mains electrical supply.

Set the multimeter to measure continuity, which is usually a little "speaker" symbol. Touch the multimeter probes together, at which point you should hear a buzz or beep, and the display should read "00.0" ohms, which means that the probe wires form a continuous circuit and have no break in them. You can now test a fuse, bulb, flex or cable by putting a multimeter probe on either end of the item. If a continuous circuit is formed and the item being tested is working, the multimeter will buzz and the display should read between 0–100 ohms. If the item being tested is broken and there isn't a continuous circuit, there will be no buzz coming from the multimeter.

Tackling emergencies

It gets more and more difficult to find reliable, skilled professionals – including electricians. The Yellow Pages are jam-packed full of electricians, offering the "best", "fastest", "cheapest" service, but when it comes to finding an electrician you trust in an emergency, you really haven't got a hope. There'll be those who can't come out, especially if it's cold and wet outside, and those who will charge a king's ransom to make it worth their while. And if you do find someone willing, you won't know if they'll do a good job if you haven't used them before.

And so, it makes a lot of good sense to know how to do a bit of basic electrical DIY, so that you can at least make sure the house is safe and has light and power.

If an electrical problem does occur – and it might happen at any time of the day or night – you must be able to assess the situation and decide on the best course of action quickly. The key steps are: make safe, identify the problem and decide on the best course of action. But long before an emergency occurs, you need to collect together some basic kit so that it can be found easily when it's needed. Your kit should consist of a large box containing a wind-up torch, a powder fire extinguisher, a collection of basic electrical tools, a key for an outdoor electricity cupboard, pencil and paper, a few old cloths, a first aid kit, telephone numbers for the electricity board, neighbours, friends, and the emergency services, perhaps a bag of your favourite nibbles and anything else that you think might come in handy. Don't bother with a candle and matches because they are dangerous.

So you are well prepared, but what do you do when the lights go down?

10 Expert Points

STEPS FOR DEALING WITH A REAL EMERGENCY:

1 FIRST THINGS FIRST
The first thing to do is to think about immediate potential dangers – anything like the baby, saucepans on the gas cooker, electric appliances, or children, elderly relatives and pets wandering about. Make sure that all cookers, fires and appliances are turned off, just in case the power comes back on during the night.

2 CHECK IF IT'S A SHARED PROBLEM
Once you are happy that everybody is safe, look out of the window to see if other houses in the neighbourhood are blacked out. If there's thunder and lightning, then the chances are that the problem is widespread and the electricity board are already aware of it. If the local area is in the dark, phone the electricity board's helpline to see if they can give you a time for the power coming back on. Let your neighbours know the situation, especially the vulnerable or elderly. Then you can go back to bed, happy in the knowledge that you've done everything you posssibly can.

3 CHECK OUT THE PROBLEM
If you are the only one in the dark, then the problem has to do with the electrics in your house. Take a torch and have a swift inspection of the house, just in case there are obvious problems such as an appliance on fire or a switch or socket shorting and flashing.

4 EXTINGUISHING ELECTRICAL FIRE
If an appliance is on fire, turn OFF the switch at the socket, pull out the plug and use the powder extinguisher to put out the fire. Cover the appliance with a damp blanket and put it out in the garden.

5 RESTORE THE POWER
Take the torch and inspect the consumer unit or fuse box. Turn OFF the main switch and check the miniature circuit breakers, fuses or switches one at a time. Replace any blown fuses, or switch on the cut-outs, as necessary.

6 REPORT FAILURE OF THE SEALED FUSE
If all the fuses and switches look fine but the power is still down, then it's possible that the electricity board's sealed fuse has blown. Report the possible failure to the electricity board, but do NOT touch the fuse!

7 WAIT UNTIL MORNING TO REPAIR LIGHTING
If, on inspection, you find that it's only a single ceiling light that is at fault, WAIT until daylight before checking the light pendant.

8 TURN OFF A FAULTY LIGHTING CIRCUIT
If, on inspection, you find that although everything else is fine, all the lighting in the house is down, the fault may be in the circuit cable. Turn OFF the lighting circuit in the consumer unit or fuse box. Then WAIT until daylight and then either make the necessary checks yourself or call in a qualified electrician to make any essential repairs.

9 TURN OFF THE MAIN POWER SUPPLY
If you replace a blown fuse or tripped switch only to find that the cut-out repeats itself, turn OFF the main power supply and inspect the system in the morning.

10 CALL THE PROFESSIONALS
If you have made all the obvious checks and are still in the dark, then WAIT until morning and call in a professional.

Rules for doing emergency repairs

Although it's good to be able to DIY, there are some times when it's best to simply to call in a professional. You still need to make basic checks to ensure that everything is under control if the lights go down. But if it's dark and raining, and you are tired or super-anxious, the best advice is to turn OFF the main power supply and wait until daylight. You must NEVER even consider trying to make repairs when you are in any way tired, ill or under the influence of drink, drugs or medication. If you think some emergency action is needed, but you feel too confused or disorientated to remember what to do or how to sort it out, call a friend or relation for help.

You must NEVER try to make repairs while you feel under pressure. For instance, perhaps the TV goes dead when the rest of the family is waiting for a favourite programme. There is a lot of pressure for you to get the TV sorted. The best advice is to try another appliance in the same socket. If that works, then the problem is with the TV. Look at the plug, check the fuse and make sure that all the terminals are tight. If the TV still doesn't work, then go and watch the programme at a friend's house. What you must NOT do is try fiddling about with the inside of the TV – at any time. Another scenario might be when the electric mower suddenly stops while you are cutting the lawn. As before, the best advice is to check another appliance in the same socket first. If that works, then look at the mower plug, check the fuse and make sure that the terminals are tight. If the mower still doesn't work, then try another cable. NEVER attempt to look at the cutters while the mower is still switched on.

Here is a list of golden rules when carrying out electrical DIY:

- **Turn off the power**
  ALWAYS switch off the circuit at the MCB (remove the fuse) and check that the circuit is DEAD before carrying out any work.
- **Unplug first**
  ALWAYS unplug an appliance before opening it up.

- **Keep your cool**
  ALWAYS be calm and methodical. Getting into a panic won't solve anything.
- **Keep your head**
  NEVER try DIY when you are in a rush, anxious, under pressure, in any way ill or under the influence of drugs, drink or strong medication.
- **Check and double-check**
  ALWAYS check and double-check that connections you have made are correct, against your wiring diagrams and the advice in this book, before turning the power back on.
- **Know your limits**
  NEVER be tempted to try a task that is beyond your skills.
- **Keep your fingers out**
  NEVER try to open a sealed appliance, especially if it is still under guarantee.
- **Water and electricity don't mix**
  NEVER work with wet fingers or in wet conditions - in the rain, on a wet lawn or in a pond.
- **Assume everything's live**
  ALWAYS proceed on the assumption that cables, flexes and terminals, even earth terminals, are live and dangerous, until you have made tests with a multimeter.
- **Working on high**
  If you need to work on something above shoulder height, make sure that you find a strong stepladder in good repair to stand on. NEVER stand on a chair. If you need to use an extending ladder, make sure someone you trust is keeping the bottom of it firmly in position so that it doesn't slide.

Emergency repairs to lighting

You have switched on the main light in a room - a ceiling pendant light - and it doesn't work.

TOOLS AND MATERIALS

- Step ladder
- A torch
- 4 mm electrician's screwdriver
- Wire strippers
- Electrical long-nose pliers
- Multimeter
- Spare bulb to match existing one
- 5-amp fuse wire or cartridge to suit
- Replacement flex to match your light fitting

Essential steps to take

1. Turn the light switch OFF.
2. Try the lights in another room.
3. If the other lights go on, get a step ladder and replace the bulb with a spare. If you don't have a spare, then borrow a bulb from another room and turn the switch back on.
4. If the light still fails, turn OFF the light switch and the main power supply. Check the MCBs/fuses. If an MCB has tripped or a fuse has blown, reset the MCB or replace the fuse (see page 157) and turn on the power.
5. Go back to the room and turn the light switch back on. If the light still doesn't work, or goes off again very quickly, turn OFF the light switch and the main power supply again and recheck the MCB or fuse. If the MCB has tripped or the fuse has blown

again, the indication is that the light pendant or the light switch is at fault. Leave the power turned OFF and return to the room.

6 With the main power supply and the light switch turned OFF, remove the light switch faceplate and check that all the terminals are tight. Replace the cover, go back to the fuse box, turn on the power and then go back to the room and turn on the switch.

7 If the light still doesn't go on, turn OFF the light switch and the main power and then go back to the room. Unscrew the ceiling rose and the bulb holder covers and check that all the connections at the terminals are good and tight. Replace the covers, fit the bulb, turn on the power, and try the light switch.

8 If the light still doesn't go on, turn OFF the light switch and the main power and then go back to the room. Unscrew the ceiling rose cover and use an electrician's screwdriver to remove the pendant completely from the holder and the flex.

9 Testing one core at a time, use the multimeter, set for continuity, to check that the cores in the flex and the cable are unbroken. Check that the bulb holder is in good order. Replace any component that is damaged. Refer to "Emergency repairs to a ceiling pendant light" on page 158. Refit the holder to the flex, pass the flex through the ceiling rose cover and screw it to the terminals, making sure the ceiling rose cover is screwed in place and the bulb fitted. Turn on the power at the fuse box and then at the switch.

10 If the light still doesn't go on and/or the fuse blows, there is probably a fault with the circuit. Refer to "Emergency repairs to a dead power circuit" on page 155. If repairing a circuit looks too daunting, call in a qualified electrician.

Failure of the hot water supply

Although most houses now have central heating that heats your water as well, there are still homes that heat their water using an electric immersion heater fitted into the hot water cylinder. Typically an immersion heater will last for many years and then fail without warning, generally when you desperately need hot water! The water heating system consists of a supply from the fuse way in your consumer unit to a timer or clock which is in turn connected to the immersion heater element in the water cylinder.

Overview

You should first of all check that the fuse or MCB has not tripped in the consumer unit. Next to go to the timer or clock and make sure that no one has been playing with it and turned it off. Lastly, you need to look at the immersion heater element itself.

Fuse or MCB tripped

It is not normal for the fuse or MCB to trip unnecessarily and usually a fault in the circuit will have caused this to happen. This could easily be as a result of the immersion heater element failing.

Repair the fuse or reset the MCB and see whether the immersion heater is now working by listening - after a few minutes you can often hear the water gently bubbling in the tank. If the element is still not working, move on to the next step (below).

Problems with the timer or clock

Timers have mechanical insides in most cases; there are small parts which eventually wear away. Most clocks have a manual override switch, which is very handy if the clock mechanism has failed. Switch the timer to ON and see if the heating element is working.

If the water is now heating, you can buy a replacement timer at your leisure. It is not very expensive to leave the immersion heater on continuously for a few days; therefore, once it is working leave the switch in the on position.

Problems with the immersion heater thermostat

Turn OFF the power at the main switch. On top of the immersion heater will be a round plastic cover held in place by a single screw or nut; remove this and lift off the cover. Inside you will see the incoming power cable connected to some terminals and at the centre a small plastic thermostat unit with a little dial on top marked in degrees. Use a small screwdriver to turn the thermostat screw right up.

Replace the cover and turn on the power. If the immersion element is now working you will need to replace the thermostat unit. Be careful to leave the immersion heater on for only about 30 minutes and then turn off the timer; if you don't do this, eventually the water will boil in the cylinder.

Failed immersion element

The element is difficult to replace and is certainly a job for a plumber since the entire hot water system will need to be drained. This is not a long job and should be completed easily within one hour once the plumber has got started.

Emergency repairs to a dead power circuit

You discover that all the sockets are dead. Everything looks fine: there's no sparking, smell, signs of burning or anything unusual, but nothing requiring power works.

TOOLS AND MATERIALS

- 4 mm electrician's screwdriver
- A torch
- 13-amp fuse
- Red/brown, black/blue and green PVC tape
- Wire strippers
- Electrician's long-nose pliers
- Multimeter

Essential steps to take

1 Go round the whole house pulling out all the plugs from the sockets.

2 Go to the consumer unit or fuse box and turn OFF the main power supply at source.

3 Check all the miniature circuit breakers (MCBs) or fuses. If one of the fuses has blown or a breaker tripped, replace the fuse wire or cartridge or turn the switch on. Turn the power back on.
4 If the fuse blows or the breaker trips instantly, then there is a problem within the circuit, so there's no need to look at any of the appliances around the house.
5 If all seems well with the circuit, go back through the rooms and try out a portable appliance in each of the sockets, one at a time. Push the appliance plug into a socket and switch it on. If it works, then try the appliance out in other sockets around the house to see if it doesn't work anywhere. If it doesn't work, then try it in another socket.
6 If you find that a single appliance causes the MCB to trip or the fuse to blow, then assume for the moment that the appliance is at fault.
7 If the circuit does go down, either from the moment power is turned on or when you plug in any of the appliances, turn OFF the main power supply and pull out all the plugs from the sockets.
8 One socket at a time, remove the faceplate and check that all the connections are tight at the terminals. Make sure that the insulation around the cores is free from damage. If there are cracks or melt holes, repair the damage with the appropriate coloured tape. With a bit of luck you will see if there is a short between two cores. Replace faceplates if they are in any way damaged, cracked or discoloured. When you have checked all the faceplates, go back to the consumer unit and switch on the power.
9 If the circuit is still down, then the likelihood is that one or other of the cables within the house is damaged. Ask yourself if you have been drilling holes in walls, banging long nails into floors or, in the case of an old house, if there is a chance that pets or rodents have chewed through exposed cables. If you think a hidden cable is damaged, either call in an electrician or refer to the appropriate section in this book.

Emergency repairs to totally dead electrics

You discover that all the sockets and all the lights are dead. Everything looks fine with no sparking, smell, signs of burning or anything unusual, but nothing works. You're not sure if there has been a power failure in your local area or if the problem is one that is affecting only you.

TOOLS AND MATERIALS

- Step ladder
- A torch
- 4 mm electrician's screwdriver
- Wire stripper
- Electrician's long-nose pliers
- Multimeter
- Cable in various sizes
- Flex in various sizes
- Replacement fuse cartridges

Essential steps to take

1 If it is dark, turn OFF the main power supply and go to bed until daylight.
2 If in the morning the whole system is still down, call on your neighbours and ask them if they are having similar problems. If there is a local power cut, the electricity board should already be at work on it. If your neighbours have power, the likelihood is that the fault is with the circuits in your house.
3 Go back home and try the lights and sockets in various rooms. If the lights and sockets are still dead, run through the checks as described on pages 132, 145 and 147.
4 If, after all the checks are done, the system is still dead, there is either a fault with the electricity board's main supply cable or with one of their sealed fuses.
5 Do NOT attempt to open sealed fuse boxes or touch main supply cables. Instead, phone the electricity board's helpline for information.

Emergency repairs when a fuse blows

You live in an old house with a fuse box rather than a consumer unit. The fuses are of the rewirable type, in which the fuse holders have two pins like a plug with a screw at each pin to hold the fuse wire and a central ceramic bridge or tube. One or other of your circuits has gone down and there doesn't seem to be any power to the lights or sockets. Everything looks fine, there is no sparking, smell, signs of burning or anything unusual, but nothing is working.

TOOLS AND MATERIALS

- Step ladder
- A torch
- 4 mm electrician's screwdriver
- Small brush, e.g. an old toothbrush
- Electrical long-nose pliers
- Multimeter
- Selection of fuse wires from 5- to 45-amp

Essential steps to take

1 Having worked through the emergency repair sequence, you realize that a circuit fuse has blown and that an appliance was at fault.
2 First it's a good idea to solve the problem at its source by throwing the faulty appliance away.
3 Switch OFF the main power supply and have a close look at the fuse box.
4 Pull out the fuse holder and take it to a table or bench in a well-lit area.
5 If you look between the pins at the central ceramic bridge, you will see that the fuse wire has melted.
6 Partially loosen the two screws and use the long-nose pliers and brush to remove all traces of the old wire.
7 Take the card of wire, unroll a length of the appropriate rating and pass the end across the bridge or through the tube and wind the end clockwise around the screw. Gently tighten the screw until the wire is gripped.
8 Pull the wire across the bridge or through the tube, not too tightly. Wind the wire around the other screw and gently tighten up. Snip the wire off from the card.

9 When you are happy that the wire is firmly gripped at the two terminals, push the holder back in place in the fuse box and switch on the power.
10 Check around the house by testing out appliances at various sockets.
11 It's a good idea to have a selection of spare holders, all ready and prepared with fuse wire of the correct rating.

Emergency repairs to a ceiling pendant light

You have just moved into a new house. All the pendant lights need replacing. The ceiling roses are fine, but the flex and bulb holders are discoloured and ugly. You want to replace them with something more elegant. This is one of those tasks when an extra pair of hands is useful - a helper to hold the steps, take the bulbs and generally pass you things will be invaluable.

TOOLS AND MATERIALS

- Step ladder
- A torch
- 4 mm electrician's screwdriver
- Small brush, e.g. an old toothbrush
- Electrician's long-nose pliers
- Side cutters
- Wire strippers
- Retractable craft knife
- The best heat-resistant bulb/lamp holders you can afford
- About 400 mm of 0.5 mm^2 twin-core and earth flex for each light
- Multimeter

Essential steps to take

1 Go round the house turning off the various light switches and then go to the consumer unit or fuse box and turn OFF the main power supply. For safety, tell others in the house what you are doing.
2 With a helper close at hand, set up the step ladder and unscrew the ceiling rose cover. Undo the two terminal screws, ease out the flex cores, unhook the brown and blue flexes from their anchors and pass everything down to your helper.
3 Move everything you have taken down to a clean, level surface and remove the bulb and the shade. Your helper can clean the bulb and shade while you proceed with the job.
4 Cut a 400 mm length of flex for each pendant that you want to replace.
5 Strip 50 mm of outer insulation from one end of each flex and 100 mm from the other end of each. Be very careful not to damage the core insulation.
6 Next, remove about 10 mm of insulation from each core, twist the wires between your fingers and bend them over in half.
7 Remove the top cover from the lamp holder and connect the 50-mm long cores to the correct terminals, brown to live, blue to neutral and green+yellow to earth (if there is one). If the lamp fitting is metal, make sure there is an earth strap connected to the fitting, or fit one.
8 When you have connected the cores to the lamp terminal, ease the blue and brown cores around their respective anchors.

9 Thread the other end of the flex first through the holder cover and then through the rose cover. Make sure that both covers have been put on the right way up.
10 Connect the cores in the flex to the correct terminals in the ceiling rose. Make sure that the screws are tight, ease the cores over the rose anchors and screw the cover into place.
11 When you are happy that all is correct, fit the shade and the bulb, and move on to the next light to repeat the process.

Emergency repairs to an appliance flex

The flex on one of your top quality appliances has been fractured, stretched, chewed by the dog, or incurred some other damage to the extent that it is less than perfect and needs replacing. Everything else about the appliance looks fine with no sparking, smell, signs of burning or anything unusual, but it doesn't work. It isn't worth trying to repair a poorer quality appliance.

Although it is possible to make temporary mends to flexes by binding the damage with PVA tape, this is a very bad and dangerous idea. It's much better to pull out the plug and cut the flex in half so that it can't be used as soon as the problem occurs. That way you will be forced to get a replacement flex.

Long lengths of flex (longer than 15 m) - such as those used for a mower - can become damaged and are usually worth repairing using a two-part connector (see page 160).

TOOLS AND MATERIALS

- 4 mm electrician's screwdriver
- Electrician's long-nose pliers
- Multimeter
- Flexible tape measure
- PVC tape
- Masking tape
- Flex to match existing flex, including braided heat-resistant sheathing
- Retractable craft knife
- Wire strippers
- Pencil and paper

Essential steps to take

1 Unplug the appliance, measure the length of flex and add 200 mm for good measure. Then cut the flex in half and take off the plug. If the flex passes through a rubber grommet at the point where it goes into the casing, unscrew the casing and carefully remove the rubber grommet. Take the grommet and part of the old flex along to your electrical supplier to get a proper match for both.
WARNING: always unplug the appliance before cutting the flex.
2 When you have obtained the replacement flex and grommet, put the appliance on a work surface at a comfortable height and have a good look at the way the flex has been fitted. Use the pencil and paper to make a diagram.
3 Undo the screws on the terminal block, remove the mechanism that grips the flex - the little bridge or

the clamp - and ease the flex out. Remove the rubber grommet. Remove the plug from the old flex.

4 Set the old flex alongside the new and label the two ends of the new flex with masking tape to indicate the plug and the appliance ends.
5 Strip back the same length of outer insulation from both ends of the new flex to match the original.
6 Clip the three cores back to the same lengths as on the old flex. On the appliance end you might find that they are all different lengths, with the earth much longer. Remove about 10 mm of insulation from each core, twist the wire between your fingers and fold it back in half.
7 Pass the appliance end of the flex through the rubber grommet, through the appliance and into the terminal block. Depending upon the design of the appliance, you might well need to fiddle about to get the grommet in place at this point.
8 Ease the flex so that the three cores are tidy and connect the green/yellow to the earth terminal on the casing, the blue to the neutral terminal and the brown to the live terminals on the block. Clamp the flex tight with the clamp provided on the appliance.
9 Having checked and double-checked the wiring against your diagram, tighten up all the screws and put the casing back together. Fit the plug on the other end of the flex.
10 Finally, plug in the appliance and test it out to establish whether it works.

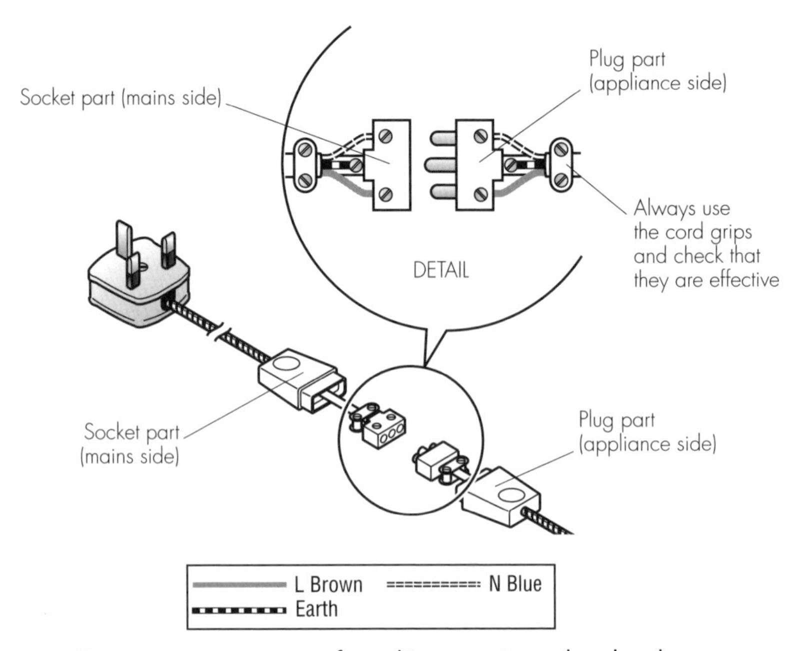

Using a two-part connector for making a repair to a long length of mower flex. Use only connectors intended for outdoor use.

Fitting a two-part connector

When you are choosing the connector make sure that you get one that is specifically designed for outside use. It must be made of rubber and it must be waterproof. When you are fitting the connector make sure that you wire the "plug" end directly to the length of flex that runs to the mower - so that the "socket" part of the connector gets to be fitted to the length of flex that runs to the wall plug. Or put another way, the order of component parts from the wall socket to the mower goes... the plug, the flex, the socket end of the connector, the plug end of the connector and then finally the length of flex that runs to the mower. Double-check when you are fitting the connector that the cord grips are securely fitted. Test the set-up by tugging the flex.

Emergency repairs to a damaged cable

One of the lighting or power cables under the floor or in the loft is damaged. Perhaps you banged a nail through a floorboard or the ceiling and into a cable. The cable is properly fitted - very tight with little or no slack. There is no other damage as far as you can see with no sparking, smell, signs of burning or anything unusual. It all looks fine, apart from the fact that you heard a bang, you can see that the cable is damaged, and the power is down. Although it is possible to make temporary mends in cables by binding the damage with PVC tape, this is most definitely a bad idea, and potentially dangerous to life and property.

TOOLS AND MATERIALS

- 4 mm electrician's screwdriver
- Electrician's long-nose pliers
- Multimeter
- Flexible tape measure
- PVC tape
- Two junction boxes
- 1 m of cable to match existing cable
- Fuse wire to match existing fuse wire
- PVC earth sleeving
- Retractable craft knife
- Wire strippers
- Hammer
- Pencil and paper
- Clips of a size to fit the cable

Essential steps to take

1 The moment you realize that there is a problem switch off the circuit at the MCB (remove the fuse) and check that the circuit is DEAD before carrying out any work. Sit down and carefully consider the implications of having the electricity turned off. For example, it might soon be dark, you might have to cook a meal for the family, or a family member may be need to work at the computer. Allow about an hour for repairing the damage.

2 Having switched the power OFF at the mains, go to the site of the damage and remove anything such as fibreglass insulation that's going to get in the way of your work.

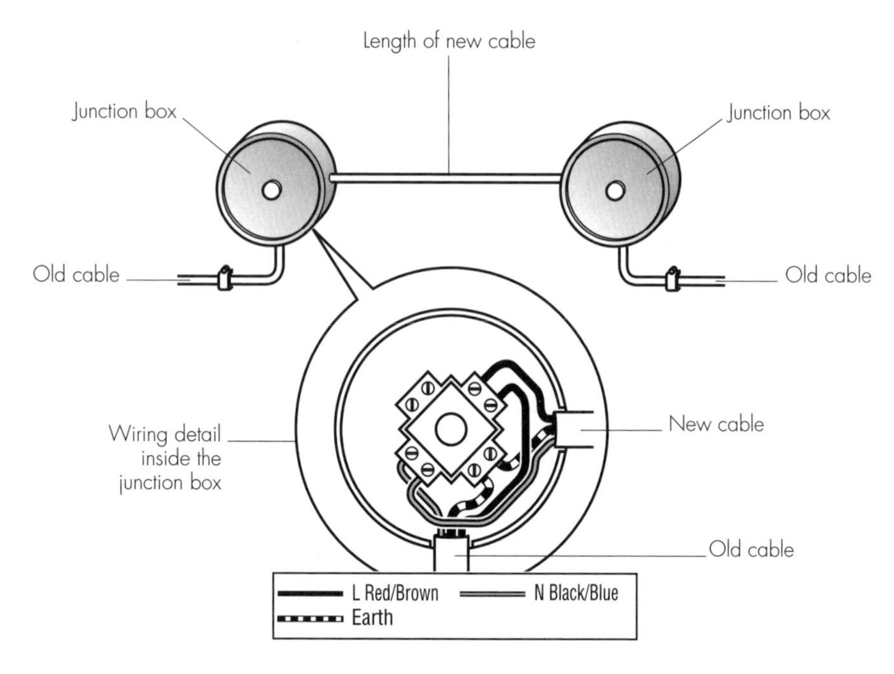

Repairing a damaged cable using two junction boxes.

3 Cut through the cable at the point of the damage so that you now have two ends. Remove any clips that are retaining the cable so that you can release the ends. You are now ready to start work on the cable.

4 Strip a piece about 50 mm in length off the outer insulation from both ends of the new cable and from both ends of the damaged cable. Then remove about 10 mm of insulation from all the cores.

5 Take the green earth sleeving and cut four 40 mm lengths. Slide the lengths of earth sleeving over the earth cores.

6 Take the two junction boxes and screw them to the joists near the ends of the damaged cable.

7 Slide the cores from one end of the prepared damaged cable into the terminals in one junction box. Repeat the procedure with the other end of the damaged cable in the other junction box.

8 Take the length of new cable and slide the cores under the correct terminals in each junction box, the red/brown live core with the red/brown from the old cable, the black/blue to the black/blue and the earth to the earth.

9 Screw the junction box lids securely in position and clip the new cable in place.

10 Replace the repaired fuse in the fuse box, switch on the power and the job is done. You can now use your appliances and lights again.

Emergency repairs to a bathroom pull-cord switch

The pull-cord switch in your bathroom has been slowly getting more and more difficult to operate. Sometimes the light comes on at the first pull, but more usually, it requires a series of sharp tugs to achieve a result. The cord has been knotted and the whole fitting looks like it is parting company from the ceiling.

TOOLS AND MATERIALS

- Step ladder
- 4 mm electrician's screwdriver
- Electrician's long-nose pliers
- Multimeter
- Retractable craft knife
- Wire strippers
- Pencil and paper
- White nylon cord to match existing cord
- Pull-cord switch

Essential steps to take

1 During daylight hours switch off the circuit at the MCB (remove the fuse) and check that the circuit is DEAD before carrying out any work.

2 Open the bathroom door and position the step ladder so that other members of the household can see what you are doing and won't trip over you as you work.

3 Unscrew the lower part of the switch and gently ease it away from the backing plate so that you can establish what the problem is.

4 Make a diagram of the wiring so that you know how the connections have been made.

5 If the string has come loose, simply pop it back into position.

6 If the backing plate is beyond repair, unscrew it and ease it away from the ceiling. Buy a new backing plate.

7 Refit a new unit in the reverse order.

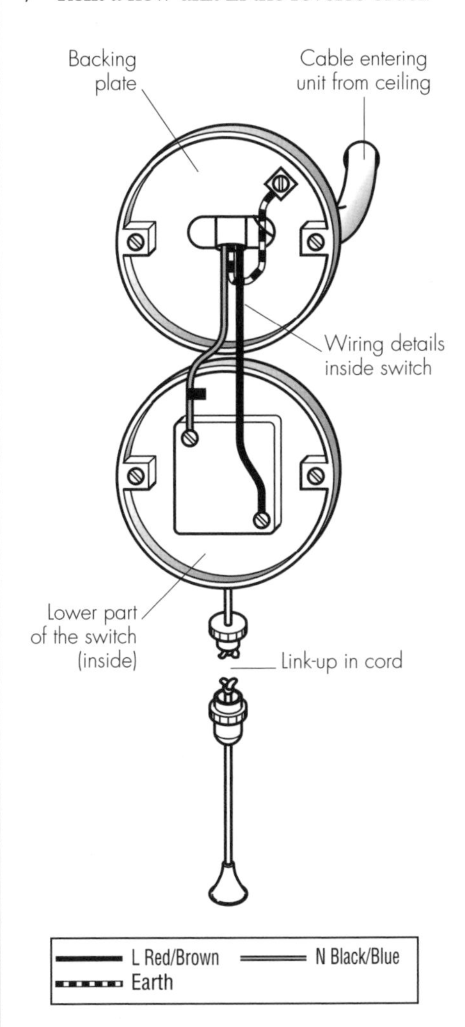

Bathroom ceiling pull-cord switch.

Glossary

Aligning
The procedure of setting one part on or against another, such as a socket on a wall, with the help of a measure and spirit level in order to obtain a good line and fit.

Alternating current
An electric current that reverses its direction many times a second at regular intervals. A multimeter averages out the actual voltage levels to give a measurement for practical purposes. In the UK, this is 230 V AC.

Amp
Abbreviation for amperage or ampere (A). The quantity of an electrical current is measured in amperes, for example the rated current of an electrical device. The amount of current used is measured in amps (A) and can be calculated by dividing the wattage (W) of the appliance by 230.

Bolster chisel
See *Floorboard chisel*.

Busbar
The brass terminal strip in a modern consumer unit.

Cable
The device providing a path along which the electricity flows. It is made up of metal conductors, or cores, covered with a protective outer sheath of plastic or PVC.

Most cable has three cores, the live sheathed in red/brown, the neutral in black/blue and the earth bare. The bare earth between the end of the cable and the fitting it is connected to must always be protected by green/yellow earth sleeving.

Cable comes in various sizes and ratings, each designed for a specific task. If in doubt about the type to use, consult a specialist supplier, detailing the work planned, and then buy the cable by the roll.

Ceiling rose
The component that links a lighting circuit and a pendent light. It can act either as a junction box in its own right in a loop-in lighting circuit or as a simple link in a junction box lighting circuit. Standard roses measure about 80 mm in diameter with a projecting depth of about 30 mm. The base plate is screwed directly to the ceiling and the cables from a loop-in or a junction box circuit, and the flex from the light fitting, are wired up to terminals on the base plate.

Some modern ceiling roses are designed so that they can be slid into place, a bit like a plug, so that you can easily remove the whole fitting for cleaning and/or fitting a new shade.

Circuit
The pathway along which electricity flows. Current flows like a stream along the live conductor to its destination - a lamp or appliance - and then back along the neutral conductor to its source.

There are power circuits, which feed appliances plugged or wired into sockets, and there are lighting circuits for wall and ceiling lights. There are two types of power circuit: a ring circuit and a radial circuit.

On a ring circuit, the pathway is 2.5 mm^2 twin-core and earth cable. The live, neutral and earth conductors in the cable run from a single 20-amp miniature circuit breaker (MCB) or fuse in the consumer unit or fuse box, through a number of socket outlets and then back to the same MCB or fuse, where they are paired with their twins - red/brown to red/brown, black/blue to black/blue, bare earth to bare earth - under the appropriate terminals. A ring circuit can be likened to a pearl necklace, in which the sockets are pearls and the MCB or fuse is the necklace fastening where the ends pair up.

On a radial circuit, the cable runs from 20-amp MCBs or fuses through five or so sockets. A radial circuit can be likened to an octopus, with the cables running out like the tentacles from the consumer unit or fuse box terminal, through a line of sockets, to finish at the last socket in the line. Each tentacle stands alone, with one end of the cable starting at its own separate MCB or fuse in the consumer unit or fuse box and the other end finishing at a socket.

Continuity tester
See *Multimeter*.

Cold chisel
A chisel used in conjunction with a club or lump hammer, to chop holes in plaster and brickwork.

Colour-coding
The conductors, or cores, in cable and flex are sheathed in coloured plastic or PVC, so that you can swiftly identify their status. Earth conductors in flexes and cables are covered with green+yellow sheathing. Neutral conductors in cable are sheathed in black/blue and those in flex are sheathed in blue. Live conductors in cable are sheathed in red/brown and those in flex are sheathed in brown.
Live = red/brown or brown
Neutral = black/blue or blue
Earth = green+yellow
A good memory aid is to think of green+yellow earth conductors as being like grass, red/brown and brown live conductors as being hot and dangerous like fire, and black/blue and blue neutral conductors as being still and neutral like darkness or water.

Conduit box
A box used at the back of some light fittings to contain the bulk of the connections. They have side and back entry holes, and fixing lugs that match up to the light fitting.

Consumer unit
A unit, sometimes known as the main fuse box, often found in the garage, under the stairs or in its own special cupboard, which contains all the switches, fuses and/or miniature circuit breakers (MCBs) governing the different circuits that run around your house. In older houses, the fuse box is usually a rather large black, brown or cream box, in which each of the circuits is protected by a rewireable fuse. When there is a problem, the fuse burns out and you have to turn OFF the main power supply, identify and pull out the offending fuse block, replace the fuse wire, solve the problem that

caused the fuse to blow and then switch the power back on.

With a modern installation, the fuse box is replaced by a neat white box complete with a row of small switches or miniature circuit breakers (MCBs). When there is a power cut, all you have to do is turn OFF the main power supply, identify and sort out the problem, and then switch on or push back the MCB.

Cross bonding
The system whereby the various high-risk metal items in your house are protected by a dedicated earth cable that links them directly to the earth terminal in the consumer unit or fuse box. For example, the earth cable might run from metal plumbing pipes under the bath to the cast-iron bath, then on to the stainless steel sink, the stove in the kitchen, the metal water tank in the loft and so on, to finish up at the consumer unit or fuse box. The earth cable is attached to each item with a clamp or clip that includes a metal tag. If there is a problem, the dangerous renegade currents are swiftly and safely run back along the dedicated earth cable to the consumer unit or fuse box.

Dimmer switch
A type of light switch that allows you to vary the intensity of the light from full brightness down to a subtle glow, a bit like turning down an oil lamp. A dimmer switch conversion kit can be fitted to an existing switch in a few minutes.

Double-pole switch
A specially designed switch to isolate and protect an appliance from the main supply by cutting or making a break at both the live and neutral sides of the circuit. This is a good and safe option for appliances, like cookers and water heaters, which are permanently connected to the mains. They come in many shapes and designs, from wall switches with red neon indicators to ceiling switches with pull-cords.

Drill and drill bit
Tools used together to make holes in timber or masonry. Two drills are needed for electrical work – a mains electric power drill for heavy-duty drilling, such as for deep holes in brickwork, and a hand or cordless drill to use when the power is switched off. A good selection of bits in a range of diameters and lengths is also needed, including a 500 mm masonry bit to drill through brick walls.

Earthing
A safety feature that protects the user by supplying an easy, safe passage for renegade currents, which would otherwise flow into a person or appliance. Every circuit supplies this passage via the green+yellow conductor that runs alongside the live and neutral. The short length of bare earth core that runs from the end of cable or flex to a switch, socket, or appliance must be protected by a length of green+yellow earth sleeving. Large high-risk metal items, like iron baths and steel sinks, must be

connected to a dedicated earth that runs back to the consumer unit or fuse box. See *Cross bonding*.

Electrical screwdriver
See *Screwdriver*.

Electrician's pliers
Pliers with insulated handles for taking a firm grip on a component. You will need long-nosed pliers for handling delicate wires and a large pair of pliers for heavy work such as twisting wires and clipping cables. It pays to get the best quality you can afford.

FCU
See *Fused connection unit.*

Faceplate
The covers seen on switches, sockets, ceiling roses etc., are known as faceplates. They need to be in good condition, completely free from cracks, tightly screwed in place and earthed if necessary. Cracked faceplates are dangerous and must be replaced.

Finishing
The procedure of tightening screws, adjusting the level of sockets and switches, sweeping up, wiping filler into holes, etc.in order to bring the job to a clean and tidy conclusion.

Flex
The flexible cord that links a light or appliance to the lighting or power circuit. Like cable, there are a number of metal cores within insulated sheaths, but there are important and characteristic differences. By definition, flex is much more flexible than cable. Some flex has no earth core as, for example, those on all plastic appliances that are double insulated. If there is an earth core in a flex, it is sheathed. Flex is round in section rather than flat.

Modern flex is also colour-coded differently from cable. The earth core is still green+yellow, but the live core is brown and the neutral is blue. Some old appliances have flexes in the old colour codes, which are red/brown for live, black/blue for neutral and green+yellow for earth.

Floorboard chisel
A wide-bladed cold chisel, sometimes called a bolster chisel, used in conjunction with a club or lump hammer, to lift floorboards and to chop holes in plaster and brickwork.

Fuse
A length of wire that is designed to melt at a set current demand and in so doing cut the circuit. Fuse holders are colour-coded for easy recognition:
white = 5-amp
blue = 15-amp
yellow = 20-amp
red = 30-amp
green = 45-amp
Always replace a fuse with one of the same rating and never use anything other than the recommended fuse wire or cartridge.

Fuse box
See *Consumer unit*.

Fused connection unit
A unit that allows a junction between a higher-rated cable and a lower-rated cable or flex, with a fuse that will blow in order to prevent the smaller cable being overloaded.

Gangs
The group of switches or sockets in an outlet. Switches are most commonly seen in one- and two-gang forms; they are also available in three, four and six gangs. A six-gang switch is about the same size as a double socket outlet.

Grommets
The little rubber or plastic rings fitted into the switch and socket cable entry holes of metal mounting boxes to protect the cable insulation from chafing.

Hacksaw
A small bow-shaped saw with a narrow blade tensioned across it. A couple of different hacksaws and a good supply of spare blades are needed - a large one for general tasks such as cutting bolts, pipes and wires, and a small junior hacksaw for cutting plug and switch cases.

Hammer
A tool for hitting in nails. Three hammers are needed - a claw hammer for general work such as banging in nails and clawing out old screws, a small tack hammer for banging in cable clips, and a large-headed club or lump hammer, used in conjunction with a flooring or bolster chisel to lift floorboards and chop holes in plaster and brickwork.

Insulation
The term for all the coverings, such as plastic, rubber, woven coating, etc., on cable and flex, which protect the live conductors and stop them being touched.

Insulation tape
Plastic PVC sticky-back tape that comes in a range of colours to match the colour-coded sheathing on cables and flexes. It is used to identify and insulate cores on switch circuits, where the bare core runs from the cable to the terminal.

Junction box
The boxes, screwed in place in the loft or ceiling space, and used on junction box lighting circuits to link the ceiling rose and the switch to the circuit cable and on power circuits to connect in spurs. Boxes for lighting circuits are usually about 80 mm in diameter and have four terminals. The screw-on cover can be rotated to create a number of cable entry options.

Lamps
Trade term for bulbs or fluorescent tubes.

Live
The term to describe a terminal, switch or electrical component on an appliance, or the cable or flex core, which carries current. Live cores are coded red/brown in cable and brown in flex.

Loop-in light circuit
A lighting circuit created by looping the cable directly from one ceiling rose to the next, rather than using junction boxes. In this system the rose becomes the junction box.

MCB
See *Miniature circuit breaker.*

Measure
A tool marked off at regular intervals for taking metric and imperial

measurements. You need two measures - a wood, plastic or metal rule for sizing and marking things like the precise size and position of holes, and a small flexible or retractable measure for general measuring such as for fixing the position of switches and sockets, and for measuring lengths of cable.

Miniature circuit breaker
The little switches, also called MCBs, in modern consumer units that fulfill the same function as a fuse. If there is a problem, the appropriate MCB trips and cuts off the power. Simpler than working with a fuse, you then solve the problem and turn the MCB switch back on.

Mounting boxes
Metal and plastic boxes that are either surface-mounted or buried in cavities and used for mounting sockets and switches on to walls. They come in all manner of shapes, sizes, and materials.

To use one, tap out selected cable entry holes, pull the cable through into the box, and screw the box in place. The metal types have plastic grommets that pop into the cable entry holes to protect the cable from coming into contact with or chafing on the metal.

Multimeter
Sometimes referred to as a continuity tester, this is a good easy-to-operate low-cost piece of battery-operated equipment used for testing cables, bulbs, fuses, etc.

To operate it, switch it on to continuity test and touch the ends of the cables or fuses to be tested with the two probes. See page 149 for more detail.

Neon tester
A screwdriver with a small neon indicator built into the handle.

To use it, hold it like a dagger with the blade touching the component to be tested while depressing the metal button on the end of the handle. If the component is live, the neon lights up.

Neutral
The term for one of the cores in cable or flex, black/blue in cable and blue in flex, which carries the current back to its source.

One-way switch
A light switch that is turned either on or off to control a single light. See also *Two-way switch*.

Pliers
See *Electrician's pliers*.

RCB
See *Residual current device*.

RCD
See *Residual current device*.

Radial circuit
A circuit made up of a number of stand-alone arms or branches. Each arm starts at a dedicated MCB or fuse in the consumer box or fuse box and then goes through a number of sockets to finish up at the last socket in the line.

Residual current device
A safety device, also known as an RCD or

residual current breaker (RCB), fitted as a permanent fixture in a circuit or as a portable item with a socket to protect you from renegade currents. If there is a problem, the device cuts the power off instantly, fast enough to protect you from harm.

Retractable craft knife
A knife with retractable, sharp blades used for all general cutting tasks.

Ring circuit
A circuit wired in a continuous loop or ring like a necklace, with both ends of the loop wired to the same terminals in the consumer unit or fuse box. As the cable runs around the ring, it feeds various sockets, with the live to live, neutral to neutral and earth to earth connections making a continuous circuit.

Saw
A long toothed blade with handles of different types for cutting timber, floorboards or plasterboard. A traditional short curved-ended handsaw designed specifically for cutting floorboards and an electric jigsaw are the most useful. The electric jigsaw is a beautifully uncomplicated, efficient, low-cost tool particularly good for cutting holes in floors and plasterboard. To use one, drill a starter hole in the board, insert the jigsaw blade into the hole so that the bed of the tool is resting flat on the board, switch on the power, and then slowly advance the tool to cut to the waste side of the required line. When you are happy with the shape of the cut, switch off the power, wait for the blade to come to a standstill and then lift it away. NEVER lift the tool away from the work piece while the blade is still in motion and always wear goggles and a facemask while using it.

Screwdriver
A tool with different blade ends to drive in matching screws. You need a good range of screwdrivers in a variety of sizes from general-purpose screwdrivers for driving in wood and masonry screws to special electrical screwdrivers. At least one of the small screwdrivers should have a magnetic tip so that you can retrieve screws and nuts from otherwise inaccessible holes. Electrician's screwdrivers should be insulated along most of their length, including the whole of the handle and the metal shank/shaft and stopping just short of the end of the blade. It pays to get the best quality you can.

Side cutters
A tool for general cutting and for stripping cable. They have the appearance of a pair of pliers, including insulated handles, and a snipping action, rather like nail clippers.

Single-pole switch
A switch, including most household light switches, which operates by cutting or connecting just the live side of the circuit.

Socket tester
A simple low-cost piece of equipment used for testing sockets. To use it, push it into the socket just like a plug and look at the

display. The little lights will instantly let you know if there is a problem or not. Testers vary, so read the manufacturer's instructions carefully.

Spirit level
A tool used for checking that fittings, fixtures and components are vertically and horizontally level or true. A small wood and brass level for checking switches and sockets, and a cheap throwaway plastic level for checking such things as cable runs, are good options.

Spur
A branch or single run of cable off a ring or radial circuit. The spur can be wired into an existing socket, or into a fused connection unit (FCU) or additional junction box.

Switch fuse unit
A small unit added to increase the capacity of a fuse box or a consumer unit.

Torch
This essential piece of equipment is a source of light when the main power is switched off. Buy a top quality torch that can be hand-held and stood on its base. Also consider a wind-up model so that you never have to think about batteries.

Trip switch
Functions in the same way as a miniature circuit breaker.

Two-way switches
Switches used in pairs that allow one light to be turned on and off from two positions, for example at either end of a hallway or staircase.

Volt
The unit of force that drives the current around the circuit. In the UK, this is usually about 230 volts (V).

Watts
A measure of electrical power that is calculated by multiplying volts by amps.

Wire strippers
A small tool, a bit like a pair of pliers, used for stripping the plastic insulation from cables and flexes. Before use, set the dial or screw to suit the core diameter of the cable or flex.

Wood chisel
A tool with a shaped metal blade for cutting wood. You need a range of sizes and they can be quite a low-cost set.

To use one, hold and guide the chisel with one hand, while tapping the handle with a hammer or mallet held in the other hand. Always wear goggles and a mask when using a chisel.

Useful contact details

Related associations

Electrical Contractors' Association
34 Palace Court
London W2 4HY
020 7313 4800
www.eca.co.uk

Energy Saving Trust
0800 512 012
www.est.org.uk

Environment Agency
National Consumer
Contact Centre
PO Box 544
Rotherham
S60 1BY
0870 8506 506

Institute of Electrical Engineers (IEE)
Maintains a list of electrician members and provides electrical safety advice
020 7240 1871

National Inspection Council for Electrical Installation Contracting
Warwick House
Houghton Hall Park
Houghton Regis
Dunstable
Bedfordshire LU5 5ZX
0870 013 0382
www.niceic.org.uk

Tools and materials suppliers

Allied Electrical
Distributors
1 Shortwood Road
Pucklechurch
Bristol
Avon BS16 9RA
01173 039 000

Ashley and Rock
Distributors
Morecambe Road
Ulverston
Cumbria
LA12 9BN

B & Q plc
Head Office
Portswood House
1 Hampshire
Corporate Park
Chandlers Ford
Eastleigh
Hampshire
SO53 3YX
0238 025 6256
www.diy.com

BEW
Distributors
Unit 4
Hanbury Road
Widford Industrial Estate
Chelmsford
Essex CM1 3AE
01245 495757
www.bewdirect.co.uk

Black and Decker
210 Bath Road
Slough
Berkshire SL1 3YD
01753 567 055
www.blackanddecker.com

Cannon Electrical Supplies
Distributors
222 Main Road
Biggin Hill
Westerham
Kent TN16 3BD
01959 570 999

Edwardes Brothers
Distributors
Units 2–6
295 Watling Street
Dartford
Kent
01322 288835
www.edwardes.co.uk

Electric Shop
Distributors
8 Acrewood Way
Hatfield Rd
St Albans
Herts AL4 0JY
0871 2884840
www.electricshop.com

Focus Do-It-All Group Ltd
Head Office
Gawsworth House
Westmere Drive
Crewe

Cheshire CW1 6XB
01270 501 555
www.focusdiy.co.uk

Homebase Ltd
Carew House
Railway Approach
Wallington
Surrey SM6 0DX
020 8835 3336
www.homebase.co.uk

Legrand Electrics Ltd
Distributors
Foster Avenue
Woodside Park
Dunstable
Bedfordshire LU5 5TA
01582 676701
www.legrand.co.uk

MEM
Distributors
Midland Electric Manufacturing
Reddings Lane
Birmingham
West Midlands B11 3EZ
0870 054 5333
www.memonline.com

MK Electric Ltd
Distributors
The Arnold Centre
Paycocke Rd
Basildon
Essex
01268 563 000
www.mkelectric.co.uk

Phase Electrical
Distributors
Unit 10 Cubitt Way
Churchfields Ind.Estate
St Leonards on Sea
East Sussex TN3 8950
01424 852 552

QVS Direct
Distributors
Unit 1
Woodgate Business Park
Vale Rise
Tonbridge
Kent TN9 1TB
01732 364 999
www.qvsdirect.co.uk

Southern Counties
Distributors
14 Harvey Road
Great Totham
Maldon
Essex CM9 8QA

Stanley UK Ltd
Europa View
Sheffield Business Park
Sheffield
Yorkshire S9 1XH
0114 244 8883
www.stanleyworks.com

Tilgear
Bridge House
69 Station Rd
Cuffley
Hertfortshire
EN6 4TG
01707 873 434

TLC Direct
Products direct
TLC Building
Fleming Way
Crawley
West Sussex RH10 9JY
01293 565630
www.tlc-direct.co.uk

Index